Digital Fire Strategies

Data, Design and Golden Thread

By
Duncan Winsbury

MAPLE
PUBLISHERS

Digital Fire Strategies

Author: Duncan Winsbury

First Published in 2026

ISBN 978-1-80723-080-7 (Paperback)
978-1-80723-081-4 (Hardback)

Book Cover Design and Layout by:
White Magic Studios
www.whitemagicstudios.co.uk

Published by:
Maple Publishers
Fairbourne Drive, Atterbury,
Milton Keynes,
MK10 9RG, UK
www.maplepublishers.com

A CIP catalogue record for this title is available from the British Library.

Preface

The fire strategy has long stood at the heart of every fire engineering project — a document that defines, justifies, and communicates how life safety objectives will be achieved. In my earlier book, *Fire Strategies – From Principles to Practice*, I explored the philosophy, structure, and professional responsibilities that underpin this essential element of building design.

Yet even as those principles remain constant, the tools and expectations that surround them are changing profoundly. The industry now stands at the threshold of a digital transformation that is reshaping how fire strategies are conceived, developed, and maintained. The shift is not merely from paper to screen; it is from static document to living data ecosystem.

Digital Fire Strategies – Data, Design, and the Golden Thread builds upon the conceptual framework established in *Fire Strategies* and extends it into the digital domain. It examines how Building Information Modelling (BIM), data interoperability, and the regulatory *Golden Thread* are converging to redefine the fire engineer's role.

This book addresses the new competencies and workflows that will shape the next generation of practice — from the integration of CFD and evacuation models within digital design platforms to the use of digital twins for live fire safety management. It also confronts the challenges of cybersecurity, data governance, and long-term stewardship of safety information.

The objective is simple but urgent: to equip fire engineers, designers, and asset managers with a clear understanding of what a *Digital Fire Strategy* truly is — not a rebranded report, but a verifiable, interoperable, and enduring record of fire safety intent across the building lifecycle.

I am deeply grateful to the many colleagues and industry professionals whose work in digital design, regulation, and building safety has inspired this transition. Their contributions mark a pivotal moment in the evolution of our discipline, one that will ultimately deliver safer, more transparent, and more accountable buildings.

Duncan Winsbury

October 2025

Introduction

The past two decades have seen an extraordinary acceleration in digital technology within the built environment. Architects, engineers, and contractors now routinely collaborate through Building Information Modelling (BIM) environments, cloud-based coordination tools, and real-time project management platforms. Yet, until recently, the fire strategy — one of the most critical documents in the design process — remained largely isolated from this transformation.

Traditionally, fire strategies have been developed as narrative documents supported by static drawings. Their content has always been rich in technical detail but poor in data connectivity. Once published, they often became detached from the live design model and rarely evolved in step with the building they served. This disconnection is increasingly untenable in a world where safety, accountability, and traceability are paramount.

The introduction of the **Building Safety Act 2022** and the requirement for a **Golden Thread of information** have fundamentally changed expectations. Fire safety data must now be accurate, accessible, and continuously maintained. The **Digital Fire Strategy** is the mechanism through which this can be achieved — a dynamic framework that embeds fire safety logic directly within the digital representation of the building.

This book explores that transformation in depth.

It begins by revisiting the principles of fire strategy development outlined in *Fire Strategies – From Principles to Practice*, reaffirming the central importance of clear design intent, defensible performance criteria, and interdisciplinary coordination. From there, it advances into the digital realm, explaining how these same principles are now being encoded within data environments and connected to the broader design ecosystem.

Through its eleven chapters, *Digital Fire Strategies* covers:

- The regulatory and technological drivers behind digitalisation.
- The structure and interoperability of fire safety data.
- The integration of modelling tools such as Smartfire and FDS within BIM environments.
- The management of fire safety information through Common Data Environments (CDEs).
- The long-term maintenance of digital strategies using digital twins, IoT sensors, and CAFM systems.
- And finally, the emerging role of artificial intelligence, predictive analytics, and automation in the fire safety lifecycle.

This is not a theoretical vision but a practical roadmap. Each chapter connects real tools and standards — BS 8644-1, ISO 19650, Autodesk BIM 360, Asite, Dalux, and others — with the daily work of fire engineers and building safety professionals. It outlines how the creation, validation, and maintenance of fire strategies can now occur within an interconnected digital framework, where every design decision is traceable, auditable, and verifiable.

The purpose of this book is twofold.

First, to help professionals navigate the technical and procedural realities of this digital transition. *Second*, to encourage the profession to embrace a new mindset — one that views the fire strategy not as a deliverable, but as a *living system of safety assurance.*

In the years ahead, the *Digital Fire Strategy* will become central to how compliance is demonstrated, how safety is managed, and how trust is maintained between designers, clients, regulators, and building occupants. It represents not the end of traditional fire engineering practice, but its logical and necessary evolution.

CONTENTS

Chapter 1
The Evolution of Fire Strategy Practice

1.1 Introduction

The fire strategy has always been a cornerstone of fire safety design — a structured framework through which the designer and engineer define how life safety, property protection, and resilience objectives are achieved. Over time, the form and content of these strategies have evolved in response to changes in legislation, technology, and professional practice.

Today, we stand at the beginning of another major shift: the transition from **narrative-based fire strategies** to **data-driven digital systems**. Understanding this evolution provides essential context for what follows, revealing how the principles of fire safety design have remained constant even as their expression and management methods have transformed.

1.2 From Prescription to Performance

In the decades following the introduction of national building codes, fire safety design was dominated by **prescriptive requirements**. Compliance was largely a matter of demonstrating adherence to specific rules and dimensions — compartment sizes, travel distances, door widths, and separation times.

This approach, while simple to apply, offered limited flexibility and did not always accommodate the complexity of modern architecture. As buildings grew taller, larger, and more sophisticated, prescriptive guidance alone could no longer guarantee safety or enable innovation.

By the late 20th century, **performance-based design** began to emerge, allowing engineers to demonstrate compliance through quantifiable outcomes rather than rigid prescriptions. This shift marked the birth of the modern fire strategy — a document that articulated the rationale behind the design, presented analytical evidence, and justified departures from conventional codes.

The introduction of **BS 7974:2001 – Application of Fire Safety Engineering Principles to the Design of Buildings** provided a structured framework for this approach in the UK. It established fire strategy development as an engineering discipline in its own right, grounded in scientific principles and professional judgement.

1.3 The Rise of the Formal Fire Strategy Document

By the early 2000s, fire strategies had become an established requirement for complex buildings. Their format, as explored in *Fire Strategies – From Principles to Practice*, typically included:

- Fire safety design philosophy
- Identification of design codes and standards
- Means of escape and evacuation strategy
- Fire detection and alarm systems
- Compartmentation and structural fire protection
- Firefighting facilities
- Management, maintenance, and evacuation procedures

These elements were presented as a structured narrative supported by drawings and schematics. The fire strategy acted as a *communication bridge* between design disciplines, regulators, and building operators.

However, despite their importance, fire strategies remained largely **document-centric**. They existed as PDFs or printed reports detached from the live design model. Revisions were often manual and changes made to the building geometry or systems were not automatically reflected within the strategy. This created the risk of divergence between the *as-designed*, *as-built*, and *as-managed* states of a building.

1.4 The Limitations of Traditional Approaches

While traditional fire strategies provided a clear record of design intent, their static nature introduced several recurring challenges:

- **Information Silos:** Fire strategy data was often isolated from architectural and engineering models, requiring duplication and manual cross-referencing.
- **Version Control Issues:** Updates during construction or operation were frequently undocumented or inconsistent.
- **Loss of Traceability:** Once submitted for approval, many strategies were never updated to reflect final as-built conditions.
- **Limited Accessibility:** Stakeholders — particularly building owners and facilities managers — often had limited visibility of the strategy's technical content or implications.
- **Reactive Management:** Fire safety issues were identified retrospectively rather than through continuous oversight.

These weaknesses became increasingly visible in complex developments, high-rise residential buildings, and projects with evolving occupancy profiles. The Grenfell Tower tragedy in 2017 further exposed the consequences of fragmented information and the absence of an enduring safety record.

The industry's response — through the **Building Safety Act 2022** — would redefine not only the regulatory framework but also the way in which fire strategies are conceived, stored, and maintained.

1.5 The Digital Imperative

The concept of the **Golden Thread of information** introduced by Dame Judith Hackitt's *Independent Review of Building Regulations and Fire Safety* represents a watershed moment. It requires that all building safety information — including the fire strategy — must be **accurate, up-to-date, and easily accessible** throughout the building's lifecycle.

This demand cannot be met through traditional document-based methods alone. It requires digital transformation.

As building design and management increasingly take place within **Building Information Modelling (BIM)** environments, the opportunity arises to embed fire strategy information directly within the digital model. Each fire-resisting element, door, detection zone, or suppression system can now carry metadata describing its fire performance, certification, and maintenance requirements.

This transforms the fire strategy from a static report into a **living digital dataset** — interconnected, traceable, and responsive to change.

1.6 Emerging Technologies and Integration

The transition toward digital fire strategies is supported by a suite of emerging technologies:

- **BIM and Common Data Environments (CDEs):** Platforms such as Autodesk BIM 360, Asite, and Dalux enable structured collaboration and controlled information exchange.
- **Computational Fluid Dynamics (CFD) and Evacuation Modelling:** Tools like Smartfire, PyroSim, and Pathfinder can now link directly to BIM geometries, closing the gap between analysis and design representation.
- **Digital Twins:** Real-time digital replicas of buildings that incorporate live data from sensors and systems.
- **Artificial Intelligence and Automation:** Algorithms capable of validating design data, predicting performance, and detecting deviations from the fire strategy intent.

Together, these technologies create the foundation for a new model of practice — one where the fire strategy exists as part of the **building's digital DNA**, continuously updated and verified against real-world data.

1.7 The Road Ahead

The evolution from prescriptive rules to digital integration represents more than a technical change; it marks a **cultural transformation** within the profession. The fire engineer of the future will not only interpret codes and conduct analysis but also manage data, configure digital workflows, and validate models within collaborative design ecosystems.

This book, *Digital Fire Strategies – Data, Design, and the Golden Thread*, charts that transition in detail. It explains how digital methods can enhance accuracy, transparency, and accountability while maintaining the rigour and engineering judgement that have always defined the discipline.

The chapters that follow examine the mechanisms of this change: how data structures are defined, how BIM integration occurs, how modelling tools can be embedded within digital environments, and how these systems can support ongoing management and compliance.

In doing so, the fire strategy is reimagined not as a static deliverable, but as a **continuous system of assurance** — the foundation of safer, smarter, and more resilient buildings.

Chapter 2
The Regulatory Landscape

2.1 Introduction

The evolution toward **digital fire strategies** is inseparable from the transformation of the **regulatory environment** governing building safety. In the United Kingdom, a succession of reviews, reforms, and new standards have reshaped expectations for how fire safety information is created, shared, and maintained.

At the centre of this change is the recognition that **information itself is a safety-critical asset**. Following the Grenfell Tower tragedy in 2017, it became clear that fragmented documentation, outdated records, and unclear lines of accountability posed significant risks to life safety. The traditional approach—where fire strategies existed as isolated reports—was no longer fit for purpose. This chapter examines the regulatory forces driving the digitisation of fire strategies, focusing on the **Building Safety Act 2022**, the **Golden Thread of information**, and supporting standards such as **BS 8644-1:2022** and **ISO 19650**. It also explores the emerging role of the **Building Safety Regulator (BSR)** and the increasing alignment of digital information management with legal accountability and competence.

2.2 The Post-Grenfell Reform Agenda

The Grenfell Tower fire represented a watershed in modern building safety regulation. Dame Judith Hackitt's *Independent Review of Building Regulations and Fire Safety* (2018) exposed systemic weaknesses—fragmented oversight, insufficient competence, and the absence of a clear, continuous record of safety-critical decisions.

Her recommendations called for:

- A new **regulatory framework** that ensures accountability throughout a building's lifecycle.
- The creation of a **Golden Thread** of building safety information.
- The appointment of clear dutyholders and the strengthening of professional competence.
- Greater use of **digital systems** to manage and retain safety information.

These principles underpin the **Building Safety Act 2022**, which forms the legal backbone for digital fire safety information management in the UK.

2.3 The Building Safety Act 2022

The Building Safety Act 2022 (BSA) represents the most substantial reform of building safety law in a generation. Its implications extend beyond high-rise residential buildings to influence the **entire construction sector**, embedding the requirement for accurate, enduring, and traceable information across all phases of a building's life.

2.3.1 Dutyholders and Accountability

The Act defines specific roles:

- **Client**
- **Principal Designer**
- **Principal Contractor**
- **Accountable Person**
- **Building Safety Manager** (in some contexts)

Each dutyholder carries responsibility for the creation, validation, and preservation of fire safety information. In practice, this means the **fire strategy** must now exist as a **living record**, continuously aligned with the as-designed, as-built, and as-managed states of the building.

2.3.2 Gateway Process

The BSA introduces three statutory "Gateways":

- **Gateway 1 – Planning Stage**: Fire safety considerations embedded early in design, supported by Fire Statements.
- **Gateway 2 – Before Construction**: Demonstration that the design meets functional fire safety requirements, including submission of the fire strategy and supporting analysis.
- **Gateway 3 – Completion**: Assurance that the built asset complies with the approved design and that all safety-critical information is handed over digitally.

The Gateway system ensures that the **fire strategy** is no longer a static submission document but a **progressive record of design intent and verification**.

2.3.3 The Golden Thread

The Act mandates that the "Golden Thread" must:

- Contain **accurate and up-to-date information**.
- Be **stored digitally** in a **secure, retrievable format**.
- Support both **design decisions** and **operational safety management**.
- Be accessible to those who need it, including the regulator and emergency services.

This legal requirement directly underpins the concept of a **Digital Fire Strategy**, positioning it as the primary repository for fire safety intent within the building's digital ecosystem.

2.4 The Building Safety Regulator (BSR)

The **Building Safety Regulator**, established within the Health and Safety Executive (HSE), oversees implementation of the new framework. Its duties include:

- Regulating higher-risk buildings.
- Overseeing competence frameworks for professionals.

- Driving cultural change and promoting transparency.
- Maintaining oversight of information integrity across the built environment.

The BSR is empowered to request access to fire safety documentation at any time. In practice, this means digital fire strategies must be **readily retrievable**, **version-controlled**, and **audit-ready** — qualities that cannot be achieved through paper-based records.

The Regulator's expectations are clear:

"Information should be structured, digital, and trustworthy — enabling decisions to be made on the basis of verified data rather than assumptions or incomplete records."

2.5 BS 8644-1:2022 – Digital Management of Fire Safety Information

Published by the British Standards Institution, **BS 8644-1** provides the operational framework for implementing the Golden Thread in relation to fire safety. It specifies how fire safety information should be **structured, classified, and exchanged** across digital systems.

Key principles include:

- Use of **consistent data schemas**, typically based on IFC (Industry Foundation Classes).
- Definition of **information requirements** across design, construction, and occupation.
- Alignment with **ISO 19650** principles for information management.
- Maintenance of a **digital fire safety file** within the Common Data Environment (CDE).

For the fire engineer, BS 8644-1 effectively transforms the fire strategy into a **structured dataset**. Each component—doors, dampers, smoke vents, detection systems—can be assigned attributes describing its fire performance, maintenance responsibility, and associated documentation.

This allows:

- Machine-readable validation of design data.
- Direct linkage between design intent and as-built verification.
- Seamless integration with CAFM (Computer-Aided Facilities Management) systems for in-use management.

2.6 ISO 19650 – Information Management Framework

The **ISO 19650** series provides the international framework for managing information over the lifecycle of a built asset using BIM. While not specific to fire safety, it defines the principles that make the **Digital Fire Strategy** possible.

Its core themes—**information requirements**, **structured exchange**, and **Common Data Environments**—ensure that every piece of fire safety data can be created, validated, and managed systematically.

In practice, ISO 19650 establishes:

- **How** information should be named, stored, and exchanged.
- **Who** is responsible for each data container.
- **When** information must be delivered and approved.

When combined with BS 8644-1, these standards create a unified framework for the **digital governance of fire safety information**, ensuring that data integrity and accessibility are embedded within every project stage.

2.7 BSI Flex 8670 and Competence

BSI Flex 8670 introduces a competence framework for those managing building safety information, including fire engineers and digital specialists. It emphasises three pillars of competence:

1. **Technical knowledge** – of fire engineering principles and regulations.
2. **Digital literacy** – understanding how data is structured, validated, and secured.

3. **Professional judgement** – interpreting information in context and maintaining integrity.

As fire strategies become digital, these competencies are essential. The modern fire engineer must now possess not only analytical skills but also the ability to **interact confidently with digital systems**, ensuring that safety-critical information is both technically accurate and properly managed.

2.8 Global Alignment and Future Harmonisation

The UK's approach to digital fire safety regulation is being closely observed internationally. Similar movements are emerging across Europe, North America, and the Asia–Pacific region, driven by shared goals of transparency, traceability, and accountability.

Initiatives such as:

- The **EU Construction Products Regulation (CPR)** revisions, introducing digital product passports.
- The **Singapore Digital Delivery Framework (IMDA, BCA)**.
- Australia's **National Building Product Register** and **Digital Twin Victoria** project.

These reflect a growing global consensus: that fire safety information must exist in **digital, interoperable, and enduring form**.

Over time, the international harmonisation of data structures (e.g. IFC 5, CEN/TC 442 BIM standards) will make it possible for digital fire strategies to be shared seamlessly across borders, manufacturers, and building portfolios.

2.9 Summary

The regulatory landscape now demands a new kind of fire strategy — one that is **dynamic, data-driven, and digitally accessible**.

The Building Safety Act provides the legal foundation; the Golden Thread defines the information principles; and BS 8644-1 and ISO 19650 provide the technical framework.

For the fire engineer, this means that compliance is no longer demonstrated solely through reports and drawings, but through **data integrity and digital verification**.

In this new environment, the **Digital Fire Strategy** becomes the mechanism through which design intent, compliance, and operational safety are united — forming a continuous, transparent chain of assurance from concept to occupation.

Chapter 3
Competence, Roles, and Responsibilities

3.1 Introduction

The creation of a **Digital Fire Strategy** demands more than new tools and technologies — it requires a transformation in the **roles, competencies, and collaborative behaviours** of those involved in the design, construction, and management of buildings.

The traditional model, in which the fire engineer produced a narrative report for submission and approval, is giving way to an integrated digital process. Information is now **co-authored, co-verified, and co-managed** across multiple professional disciplines, each with defined responsibilities under law.

This chapter examines those responsibilities in the context of the **Building Safety Act 2022**, the **Golden Thread**, and the associated standards governing competence — particularly **BSI Flex 8670** and guidance from the **Building Safety Regulator (BSR)**. It also explores how the digital environment demands new forms of technical and data literacy among fire engineers, architects, and project leaders.

3.2 The Reframed Concept of Competence

Historically, professional competence in fire engineering has been defined by technical knowledge — understanding fire dynamics, codes of practice, and the application of engineering judgement. While this remains essential, the digital age introduces two additional dimensions:

1. **Digital competence:** The ability to structure, manage, and interpret digital data within collaborative systems such as BIM, Common Data Environments (CDEs), and digital twins.

2. **Information management competence**: The capacity to maintain data integrity, ensure version control, and safeguard information throughout a building's lifecycle.

In the words of the BSR:

"Competence must now include the capability to create, manage, and interpret digital safety information with the same rigour that has long been applied to physical design."

This expanded definition places new expectations on fire engineers — not only to design for safety, but also to ensure that the safety information itself is correct, structured, and durable.

3.3 The Principal Designer

Under the Building Safety Act 2022, the **Principal Designer** plays a central coordinating role. Their duty extends beyond architectural coordination to encompass **information governance** for building safety.

In digital practice, the Principal Designer must:

- Ensure that the fire strategy and associated models align with the overall design intent.
- Coordinate the **information exchange** between disciplines, ensuring that the fire engineer's data is embedded within the BIM model.
- Verify that **Gateway submissions** (particularly Gateway 2) are complete, accurate, and digitally structured
- Liaise with the **Accountable Person** to ensure that all fire safety data is transferred seamlessly into operational management systems.

In essence, the Principal Designer becomes the **custodian of the digital safety narrative** during the design and construction stages — a role that bridges technical and managerial domains.

3.4 The Fire Engineer

The **fire engineer** remains the primary author of the fire strategy, but the nature of authorship is changing. No longer confined to written documents, the fire engineer must now embed fire safety intent directly within **digital models and data environments**.

Key responsibilities include:

- Defining the **fire safety data requirements** for each project stage.
- Populating BIM elements with relevant attributes (e.g., fire resistance, reaction-to-fire rating, certification).
- Conducting performance-based assessments (e.g., CFD, evacuation modelling) and linking the results to the model.
- Ensuring that design changes are automatically reflected within the **Digital Fire Strategy dataset**.
- Validating digital information for accuracy before submission to the BSR or Building Control.
- Contributing to the **digital handover file** that forms part of the Golden Thread.

In the digital era, the fire engineer's deliverable is no longer a static report but a **multi-format package** — combining narrative explanation, visual documentation, and structured datasets.

As explored later in this book, the emerging use of **AI-assisted validation tools** and **automated compliance checking** will further extend this role, allowing fire engineers to engage directly with data-driven design review systems.

3.5 The Principal Contractor

The **Principal Contractor** carries the duty of ensuring that the fire safety design intent is realised accurately on site. Within a digital framework, this responsibility extends into **data verification and record management**.

Principal Contractors must:

- Confirm that installed fire protection systems match the **approved digital specifications**.
- Upload installation data, commissioning certificates, and photographic evidence into the project's CDE.
- Report deviations and coordinate revisions to the Digital Fire Strategy dataset.
- Collaborate with the fire engineer to validate "as-built" fire safety performance.

This process ensures that the **as-built model** reflects the true installed condition — an essential requirement for Gateway 3 compliance and for the operational phase of the building.

3.6 The Accountable Person and Building Safety Manager

Once the building is occupied, responsibility transfers to the **Accountable Person** (and, in some cases, the **Building Safety Manager**). Their duty is to maintain the building's safety information and ensure that fire risk is managed effectively throughout its lifecycle.

In digital terms, this means:

- *Maintaining the **Golden Thread** as a continuously updated digital record.*
- *Ensuring that the fire strategy, maintenance records, and inspection results remain consistent.*
- *Integrating fire safety data into CAFM and digital twin systems.*
- *Facilitating regulatory access to verified information when required by the BSR.*

Whereas the fire strategy was once viewed as a design-stage document, it is now a **permanent management asset**. The Accountable Person becomes the long-term custodian of that data, supported by facility and maintenance teams.

3.7 The Building Safety Regulator (BSR)

The **BSR** plays a pivotal oversight role in ensuring that dutyholders meet their obligations. Its remit extends across both **competence** and **information integrity**.

Key functions include:

- Defining the competence standards for Principal Designers, Contractors, and Fire Engineers.
- Setting expectations for digital information management, structure, and traceability.
- Reviewing Gateway submissions to ensure that digital fire safety data aligns with approved design intent.
- Conducting audits to verify that the Golden Thread remains complete and accessible.

The BSR's guidance increasingly emphasises the need for **standardised information formats**, automated data validation, and accessible, up-to-date digital records. In this context, the **Digital Fire Strategy** becomes the principal artefact through which compliance and accountability are demonstrated.

3.8 Interdisciplinary Collaboration

Digitalisation demands that collaboration move beyond traditional document exchange. In a BIM environment, fire engineers, architects, structural engineers, and M&E designers all contribute to a **shared digital model**.

Effective coordination relies on:

- Clearly defined **information exchange protocols**.
- Agreed **model ownership** and version control procedures.
- Regular **clash detection** and **data validation** exercises.
- Common understanding of fire safety parameters embedded within the model.

For example:

- The architect defines compartment boundaries.

- The structural engineer specifies fire resistance periods.
- The M&E designer coordinates smoke control systems.
- The fire engineer verifies compliance and uploads supporting evidence.

This collective process forms the backbone of the **Digital Fire Strategy** — a living record that integrates every element of design and construction relevant to fire safety.

3.9 Digital Competence Framework (BSI Flex 8670)

BSI Flex 8670 defines competence in terms of three overarching categories:

Competence Domain	Description	Application to Fire Strategy Practice
Knowledge	Technical, regulatory, and contextual understanding of building safety.	Fire engineering principles, codes, and analytical methods.
Skills	The ability to apply knowledge effectively to deliver safe outcomes.	Design coordination, modelling, and system verification.
Behaviour	Professional conduct, communication, and ethical responsibility.	Maintaining integrity of information, transparency, and collaboration.

In the context of **Digital Fire Strategies**, each domain extends to include digital literacy:

- Understanding data structures, metadata, and model attributes.
- Using collaborative digital platforms and CDEs.
- Managing information securely and ethically.

Competence is therefore both a technical and a **data management responsibility**, reflecting the hybrid nature of modern fire engineering practice.

3.10 Emerging Professional Roles

The digitalisation of fire safety has also given rise to new interdisciplinary roles, including:

- **Digital Fire Engineer** – a practitioner specialising in integrating fire safety data within BIM and digital twins.
- **Information Manager (Fire Safety Focus)** – a data professional ensuring compliance with BS 8644-1 structures.
- **Fire Safety Data Coordinator** – responsible for quality assurance and Golden Thread updates during operation.

These roles reflect the growing recognition that fire safety is no longer confined to design but is an **information lifecycle discipline** — spanning creation, verification, handover, and ongoing management.

3.11 Summary

The move toward **Digital Fire Strategies** is as much a human and organisational challenge as it is a technological one. The roles of fire engineer, designer, contractor, and building operator are being redefined around shared digital responsibility and continuous accountability.

Competence now encompasses:

- **Technical expertise** in fire engineering design and analysis.
- **Digital literacy** in information management and data systems.
- **Collaborative capability** across multiple disciplines and project stages.

Together, these elements form the foundation of a new professional culture — one in which safety, transparency, and information integrity are inseparable.

As the following chapters will show, this culture is enabled by structured data frameworks, BIM integration, and the development of **Digital Fire Strategy workflows** that support the entire lifecycle of a building.

Chapter 4
Structuring Fire Safety Information

4.1 Introduction

At the heart of every **Digital Fire Strategy** lies structured information. Where once a fire strategy comprised descriptive text and annotated plans, the digital approach replaces narrative with *data* — information that is organised, machine-readable, and linked directly to the physical and functional elements of the building.

This transformation enables the fire strategy to move beyond being a static report and become a **dynamic digital model**, capable of validation, version control, and long-term management.

This chapter explores how fire safety information is structured, classified, and stored in accordance with **BS 8644-1**, **ISO 19650**, and **IFC (Industry Foundation Classes)** standards. It also outlines how metadata, object attributes, and digital containers together form the foundation of the Digital Fire Strategy.

4.2 From Narrative to Data Structure

In traditional practice, a fire strategy report might describe a wall as "a 60-minute fire-resisting compartment wall constructed from two layers of 15 mm Type F plasterboard on each side of a 70 mm steel stud."

In a digital environment, this description is replaced by a structured data entry:

Attribute	Value
Element Type	Internal Partition
Fire Resistance (Integrity)	60 minutes
Fire Resistance (Insulation)	60 minutes
Construction Type	Stud Partition
Surface Finish	2 × 15 mm Type F Gypsum Board
Certification Reference	EN 1364-1 / Test Report No. 12345
Maintenance Interval	12 months
Responsible Party	Fire Engineer / Contractor

This structured approach allows:

- **Automated validation** (e.g., confirming compliance with design intent).
- **Traceability** of certification and product data.
- **Efficient updates** if performance specifications change.
- **Seamless export** to other platforms (BIM, CAFM, or regulatory portals).

Each data entry becomes a *digital record* of a safety-critical decision.

4.3 The Role of Metadata

Metadata is "data about data" — contextual information that defines how each item should be interpreted, categorised, and managed.

In a Digital Fire Strategy, metadata gives every component a digital identity, linking it to performance requirements, responsible persons, and lifecycle status.

Typical metadata for a fire-safety-critical element includes:

- **Unique Identifier (GUID):** a globally unique code for tracking through all project stages.
- **Element Classification:** e.g., wall, door, damper, compartment boundary.

- **Fire Rating Attributes:** resistance, integrity, insulation.
- **System Associations:** linked alarms, smoke control systems, or suppression networks.
- **Document Links:** test reports, product certificates, or maintenance instructions.
- **Lifecycle Status:** design / as-built / in-use / decommissioned.
- **Owner / Responsible Party:** identifies accountability at each stage.

By embedding metadata within BIM objects, the fire strategy becomes self-referential and self-auditing — a **digital safety ledger** rather than a narrative description.

4.4 BS 8644-1:2022 Framework

BS 8644-1 provides the practical framework for **digitally managing fire safety information**. It defines how data should be structured, categorised, and transferred across project stages.

Key provisions include:

- **Information Containers** – Digital "packages" of fire safety data (e.g., compartmentation, means of escape, active systems).
- **Data Classification** – Aligns with **Uniclass 2015** and **IFC** object hierarchies.
- **Attribute Sets** – Specifies the properties to be recorded for each object (e.g., fire rating, certification reference, manufacturer).
- **Change Management** – Mandates version control and audit trails.
- **Common Data Environment (CDE)** – The digital repository through which all fire safety information is shared and maintained.

Under BS 8644-1, each stage of the building lifecycle — **Design**, **Construction**, **Handover**, and **Operation** — requires the creation or refinement of specific fire safety information containers.

Lifecycle Stage	Fire Safety Information Container	Purpose
Design	Fire Strategy Model (FSM)	Defines design intent and performance objectives.
Construction	Fire Safety Record (FSR)	Captures installation evidence and test certificates.
Handover	Digital Fire Safety File (DFSF)	Consolidates verified data for operational use.
Operation	Golden Thread Record (GTR)	Maintains live, updated safety data through the building's life.

Together, these containers form the **Digital Fire Strategy System**, evolving with the project from concept to occupancy.

4.5 The IFC Data Model

Industry Foundation Classes (IFC) is the global open standard for exchanging BIM data. It enables different software platforms — such as Autodesk Revit, Graphisoft Archicad, Bentley OpenBuildings, or Smartfire CFD — to exchange consistent fire safety information.

Key IFC entities relevant to fire strategies include:

IFC Entity	Typical Application
IfcWall, IfcWallStandardCase	Fire-resisting and compartment walls
IfcDoor, IfcWindow	Fire doors, glazed screens
IfcCovering	Finishes and protective linings
IfcBuildingStorey / IfcSpace	Compartment and zone definition
IfcFireSuppressionTerminal	Sprinkler heads, nozzles
IfcAlarm / IfcSensor	Detection and alarm devices
IfcPropertySet	Fire rating, smoke resistance, certification data

By assigning fire safety properties (within IfcPropertySet), the Digital Fire Strategy can store performance criteria directly in the model geometry, ensuring interoperability between design tools and asset management systems.

4.6 Common Data Environment (CDE)

A **Common Data Environment** acts as the central digital repository for all fire safety information. It ensures that data is:

- **Authorised:** created or edited only by permitted contributors.
- **Validated:** checked for completeness and accuracy.
- **Versioned:** every update recorded with time, author, and reason.
- **Accessible:** structured for retrieval by designers, contractors, and regulators.

Leading CDE platforms include **Autodesk BIM 360 / ACC**, **Asite**, **Dalux**, and **Viewpoint for Projects**.

Within these environments, fire safety information containers defined by BS 8644-1 are maintained as linked datasets — for example, compartmentation plans linked to fire doors, detectors, and suppression zones.

The CDE therefore becomes the **digital home of the fire strategy**, accessible throughout design, construction, and operation.

4.7 Fire Safety Attribute Taxonomy

To ensure consistency, each object in the fire strategy model must carry a defined set of attributes.

A typical taxonomy might include:

Category	Attribute	Example Value
Identification	GUID	8F2C-4A11-B78D-0091
Classification	Uniclass 2015 Code	Ss_25_10_60_25 (Fire-stopping systems)
Performance	Fire Resistance	EI 60 / E 60
Compliance	Test Standard	BS EN 1364-1:2015
Certification	Product Certificate	IFCC 1829 / EXOVA Ref 12345
Maintenance	Interval	12 Months
Responsible Party	Role	Fire Engineer / Building Safety Manager

Such taxonomies enable **automated auditing**, ensuring that all safety-critical attributes are present and consistent across the model.

4.8 Linking Documents, Models, and Records

A Digital Fire Strategy is not limited to geometric models. It also links to supporting documents:

- Fire strategy narrative (PDF or structured XML).
- CFD and evacuation modelling outputs.
- Test certificates and maintenance manuals.
- Gateway submission records and approvals.

These are connected through **unique reference IDs**, allowing a user to trace any component — such as a fire door — back to its certification, specification, and maintenance history. This principle of **data linkage** transforms traditional document management into **digital traceability**, a key element of the Golden Thread.

4.9 Lifecycle Information Flow

The journey of fire safety information through a project can be visualised as:

- **Design Stage:** Fire strategy intent defined and encoded in BIM.
- **Construction Stage:** Installation data and commissioning results appended.
- **Handover:** Verified digital dataset compiled into the Digital Fire Safety File.
- **Operation:** Dataset integrated into CAFM or Digital Twin for ongoing management.
- **Refurbishment or Alteration:** Changes recorded, dataset versioned and re-verified.

This continuous flow ensures that the Digital Fire Strategy remains an **active, living system** throughout the building lifecycle.

4.10 Data Validation and Quality Assurance

To maintain integrity, digital fire safety information must undergo regular validation. This may include:

- **Automated rule-based checks** (e.g., Solibri, Navisworks Manage) verifying that every compartment boundary has a rated door or damper.
- **Manual peer review** by fire engineers and information managers.
- **Cross-disciplinary audits** before Gateway submissions.

Quality assurance protocols should define:

- Data ownership and approval workflows.
- Review frequencies and validation metrics.
- Archiving and backup requirements.

Without structured validation, the benefits of digitalisation can be undermined by incomplete or inaccurate data.

4.11 Security and Ethical Considerations

Digital fire safety data is sensitive. Plans and specifications reveal the means of escape, firefighting access, and systems designed to protect life — information that must be protected from misuse. Therefore, information security is an essential part of structuring digital fire strategies. This includes:

- **Access control:** role-based permissions within CDEs.
- **Encryption and secure storage:** especially for cloud-based systems.
- **Audit logging:** to track who accessed or modified information.
- **Compliance with data protection legislation**, including GDPR.

Ethically, professionals must also consider data accuracy and transparency. A mis-classified or outdated record could compromise safety — reinforcing the principle that **digital competence equals professional responsibility**.

4.12 Summary

The structure of fire safety information determines its reliability, usability, and value. By adopting frameworks such as **BS 8644-1**, **ISO 19650**, and **IFC**, fire engineers and designers can transform static reports into dynamic, interconnected datasets.

A well-structured Digital Fire Strategy:

- Maintains a single source of truth for all fire safety decisions.
- Supports validation, audit, and change control.
- Integrates seamlessly with design, construction, and operational systems.
- Provides the foundation for the **Golden Thread** — a continuous record of safety from concept to decommissioning.

In the chapters that follow, this structured foundation becomes the platform upon which advanced applications are built — including **BIM integration**, **CFD and evacuation modelling**, and **automated validation** — all of which define the next stage in the evolution of fire safety design.

Chapter 5
BIM Integration and Workflow

5.1 Introduction

The integration of the **Digital Fire Strategy** into **Building Information Modelling (BIM)** environments represents the practical realisation of digital fire safety. It transforms the fire strategy from a standalone document into a living component of the building model — directly linked to geometry, systems, and performance data.

Through BIM, fire safety design becomes fully embedded within the **digital design and coordination process**, ensuring that every door, wall, shaft, and escape route carries its correct fire performance attributes.

This chapter explores how BIM platforms support Digital Fire Strategies, how information flows between stakeholders, and how model-based coordination enhances accuracy, traceability, and compliance.

5.2 The Role of BIM in Fire Safety Design

BIM provides a **shared digital representation** of the physical and functional characteristics of a building. For fire safety, this allows all elements influencing protection and evacuation to be defined, visualised, and validated in one environment. Typical fire safety components represented in BIM include:

- Compartmentation and fire-resisting elements
- Fire doors and glazed screens
- Fire detection and alarm zones
- Smoke control and ventilation systems
- Sprinklers and suppression networks
- Firefighting shafts, risers, and access routes
- Means of escape, stair cores, and refuge points

Each of these components can carry **metadata** aligned to BS 8644-1 (as detailed in Chapter 4), forming part of the **Digital Fire Strategy dataset**. This integration ensures that the fire engineer's design intent is **visible, verifiable, and traceable** throughout the design, construction, and operational phases.

5.3 Fire Strategy Layers within BIM

Fire-related information can be represented within BIM models using **dedicated layers or worksets**, allowing the fire engineer and project team to visualise and manage the fire strategy independently from other disciplines.

Typical layers might include:

Layer / Workset	Content	Purpose
FS_ Compartmentation	Fire-resisting walls, floors, and boundaries	Defines separation and containment strategy
FS_Doors_Protected	Fire doors and glazed partitions	Identifies openings requiring resistance ratings
FS_Means_Escape	Escape routes, travel distances, and exits	Supports evacuation design validation
FS_Detection	Alarm devices and detector coverage	Links to zone and control data
FS_Firefighting	Shafts, risers, hydrants, and access	Visualises firefighting facilities
FS_Suppression	Sprinklers, misting systems, suppression zones	Coordinates hydraulic and zoning layouts

By isolating these layers, fire engineers can coordinate visually with architects and M&E designers, ensuring that fire safety provisions remain coherent and compliant across all design revisions.

5.4 Coordination between Disciplines

Effective integration of the Digital Fire Strategy depends on seamless **interdisciplinary coordination**. BIM provides the platform for this collaboration through model federation and shared data protocols.

Key coordination mechanisms:

- **Model Federation:** Architectural, structural, M&E, and fire models combined into a single coordinated view (via Navisworks or Solibri).
- **Clash Detection:** Automated identification of conflicts (e.g., dampers misaligned with walls, ducts breaching fire compartments).
- **Information Exchange Protocols:** Defined in the **BIM Execution Plan (BEP)** to govern data ownership, naming conventions, and update frequency.
- **Review Workflows:** Regular coordination meetings and issue tracking via CDE tools such as BIM 360, Asite, or Dalux.

Fire engineers should be active participants in this coordination process — not merely recipients of models, but contributors to data integrity and system validation.

5.5 Embedding Fire Strategy Data in BIM Elements

Every fire-critical element in the BIM model should include **property sets** containing fire safety data. These can be defined using **IFC property sets (IfcPropertySet)** or **Revit shared parameters**.

For example, a fire door might include:

Parameter	Value
FireRating	60 minutes (E60)
SmokeControl	Yes
CertificationReference	IFCC 2934
InstallationDate	2025-03-01
MaintenanceInterval	12 months
ResponsibleParty	Accountable Person – Operations
GatewayReference	G2/FD/3.4

Such structured parameters ensure that fire safety data remains attached to each physical object and automatically transfers between design, construction, and operational models.

This creates **true data continuity** — the essence of the Golden Thread.

5.6 Autodesk Revit and Fire Engineering Applications

Autodesk Revit is the most widely used authoring tool for fire strategy integration. Fire engineers can use Revit to:

- Apply fire performance parameters to walls, floors, and doors.
- Generate fire compartmentation plans automatically.
- Validate travel distances using view filters or third-party plug-ins.
- Export IFC or COBie data for downstream systems.
- Generate colour-coded fire zone diagrams for Gateway submissions.

Revit add-ins (such as **Ideate BIMLink** or **Solibri Model Checker**) allow engineers to extract fire safety data directly into spreadsheets or databases, enabling validation against design requirements and regulatory benchmarks.

5.7 Navisworks and Solibri: Review and Validation

Navisworks Manage and **Solibri Office** are powerful coordination tools for model review and rule-based validation.

- **Navisworks Manage:** Enables visual inspection and automated clash detection between architectural and fire protection elements. Engineers can assign review status (approved, revision required, non-compliant).
- **Solibri Model Checker:** Supports rule-based analysis — e.g., "Verify all fire doors in compartment walls have a minimum E30 rating," or "Identify any ducts passing through protected shafts without dampers."

These tools provide the means to perform **digital quality assurance** before submission to the regulator or client, replacing traditional mark-up drawings and narrative reviews.

5.8 BIM 360 / Autodesk Construction Cloud (ACC)

BIM 360 and **Autodesk Construction Cloud (ACC)** act as the **Common Data Environment (CDE)** where all project information — including the Digital Fire Strategy — is stored, reviewed, and approved.

Within ACC:

- The fire engineer can upload fire safety models and link supporting documents (CFD results, risk assessments, Gateway forms).
- Reviewers can comment, mark-up, and request revisions directly in the 3D environment.
- Version control ensures that each revision of the fire strategy model is archived and timestamped.
- Approval workflows confirm when the model is formally accepted into the Golden Thread record.

Integration with Autodesk's **Forge API** allows advanced functionality such as automated compliance checking and real-time linkage to Digital Twins, discussed later in Chapter 10.

5.9 Workflow Example: From Design to Digital Record

A simplified Digital Fire Strategy workflow within BIM can be illustrated as follows:

1. Model Creation (Revit / Archicad)
    a. Fire strategy elements modelled with property data.
    b. Initial fire compartments, doors, and routes defined.
2. Model Coordination (Navisworks / Solibri)
    a. Combined model checked for conflicts and omissions.
    b. Fire safety attributes validated.

3. Information Exchange (BIM 360 / CDE)
    a. Updated model uploaded to Common Data Environment.
    b. Gateway 2 submission generated automatically.
4. Approval and Handover
    a. Building Control / BSR review the model and digital fire strategy data.
    b. Approved version archived into Golden Thread repository.
5. Operational Integration
    a. Fire strategy data transferred to CAFM or Digital Twin for ongoing use.

This closed-loop workflow replaces static report delivery with a **continuous data stream**, ensuring transparency, consistency, and accountability across all project stages.

5.10 Model-Based Gateway Submissions

Under the **Building Safety Act 2022**, Gateway submissions (particularly Gateways 2 and 3) increasingly rely on digital evidence rather than paper-based documentation. The **Digital Fire Strategy model** forms the central submission artefact, supplemented by structured narrative data (PDF/XML).

Key submission elements include:

- Fire compartmentation model (IFC format).
- Schedule of fire doors, dampers, and protection systems with certification data.
- CFD and evacuation simulation results (linked through model references).
- Design change log and version history.
- Validation report confirming compliance with design standards.

These submissions allow regulators to **interrogate the model directly**, improving transparency and reducing approval times.

5.11 Collaboration Platforms and Issue Tracking

Modern collaboration environments provide real-time communication channels for managing fire safety issues:

- **BIM Track**, **Revizto**, or **Asite Field** allow users to raise, assign, and resolve issues directly in 3D context.
- Each issue (e.g., missing damper, incorrect fire rating) is linked to a model element and recorded in the CDE.
- Resolution history is automatically captured, creating a **digital audit trail** for the Golden Thread.

This approach removes ambiguity, as every design decision or change request is transparently recorded, timestamped, and traceable to a responsible party.

5.12 Integration with CFD and Evacuation Modelling

As explored further in Chapter 6, BIM models can interface with **CFD and evacuation software** such as **Smartfire**, **FDS**, **Exodus** and **Pathfinder**. Geometry is exported directly via IFC or gbXML, while results — smoke spread, temperature, tenability — can be reimported into BIM for visualisation.

This enables:

- Validation of compartmentation and ventilation design.
- Enhanced visual communication with clients and regulators.
- A permanent link between analytical evidence and the fire strategy model.

The result is a **multi-dimensional representation** of the fire strategy — geometric, performance-based, and data-driven.

5.13 Challenges and Best Practice

While BIM integration offers significant advantages, several challenges must be addressed:

Challenge	Best Practice Response
Inconsistent model standards	Define parameters in a BIM Execution Plan (BEP) at project outset.
Lack of digital competence	Provide structured training in data management for fire engineers.
Poor coordination between disciplines	Use regular federated model reviews and shared issue tracking.
Model over-complexity	Apply model filtering — fire safety data only where relevant.
Version confusion	Enforce strict version control within the CDE and maintain a model register.

Adopting these principles ensures that digital fire strategies remain robust, accurate, and compliant throughout the design process.

5.14 The Benefits of BIM-Integrated Fire Strategies

The integration of fire strategies within BIM environments offers measurable benefits:

- **Accuracy:** Reduces manual errors and inconsistencies between drawings and reports.
- **Transparency:** Allows all stakeholders to view the fire safety design in 3D.
- **Efficiency:** Simplifies coordination, reviews, and regulatory approvals.
- **Traceability:** Maintains an unbroken record of design intent and decisions.
- **Lifecycle Continuity:** Enables direct handover to operational systems.

Ultimately, BIM integration ensures that the fire strategy is **not an attachment to the design** but a **core component of it** — continuously updated, validated, and accessible to all authorised parties.

5.15 Summary

The **Digital Fire Strategy** achieves its full potential when integrated within BIM environments. Through structured data, model-based collaboration, and controlled workflows, fire safety information becomes an active part of the building's digital ecosystem.

BIM integration allows:

- Fire safety intent to be visualised and verified in 3D.
- Data to be structured, validated, and linked to physical components.
- Regulatory submissions to be simplified and transparent.
- Long-term maintenance of the Golden Thread through accurate as-built information.

In the chapters that follow, we explore how this integrated model is extended into performance-based validation — using **CFD and evacuation modelling (Chapter 6)** — and how these analyses become part of the **living digital record** that defines the next generation of fire safety management.

Chapter 6
CFD, Evacuation, and Performance Modelling

6.1 Introduction

Computational modelling has long played a vital role in performance-based fire engineering. Tools such as **Smartfire**, **PyroSim**, and **Fire Dynamics Simulator (FDS)** have enabled engineers to quantify smoke movement, heat transfer, and tenability; while **evacuation simulators** like **Pathfinder** and **BuildingEXODUS** have provided insights into occupant behaviour and egress performance.

Historically, these models existed in isolation from the building's design environment. Geometry was imported manually, results were exported as static images, and the link between the model and the built asset was weak.

In the context of **Digital Fire Strategies**, this separation is no longer sufficient.

Modelling data must now integrate seamlessly with the **BIM model**, **Common Data Environment (CDE)**, and ultimately, the **Golden Thread** — ensuring that analytical evidence is directly traceable to design decisions and remains accessible throughout the building's lifecycle.

This chapter explores how this integration is being achieved, how modelling data is structured for reuse and validation, and how computational analysis is evolving toward real-time feedback within digital twin environments.

6.2 The Role of Modelling in Digital Fire Strategies

A Digital Fire Strategy is both a descriptive and analytical construct. While its structure (as covered in Chapter 4) defines what the fire safety provisions *are*, performance modelling demonstrates *how* and *why* they will perform as intended.

Integrating CFD and evacuation simulations within the digital fire strategy offers:

- **Quantitative justification** for fire safety design assumptions.
- **Visual communication** of complex smoke and evacuation behaviours.
- **Verification evidence** for regulators and clients at Gateway 2.
- **Predictive insight** for future system optimisation.

In digital form, these models become more than one-off calculations — they form an evolving, verifiable dataset that supports **continuous assurance** across the building's lifecycle.

6.3 Computational Fluid Dynamics (CFD) Integration

6.3.1 Overview of CFD Tools

Commonly used fire CFD tools include:

- **Smartfire** – developed by the Fire Safety Engineering Group (FSEG), University of Greenwich; purpose-built for fire and smoke simulation.
- **PyroSim / FDS** – widely used for design-based modelling, supported by NIST.
- **FireFOAM / OpenFOAM** – open-source CFD platforms used in research and advanced consultancy.

Each tool simulates fire growth, smoke movement, heat transfer, and ventilation performance within a defined geometry.

6.3.2 Linking CFD to BIM Geometry

Traditionally, engineers recreated building geometry manually in the CFD pre-processor — a time-consuming and error-prone process.

Integration with BIM changes this.

- **IFC and gbXML** export functions from Revit or Archicad allow direct transfer of building geometry, materials, and boundary conditions.
- **Smartfire's BIM Importer** can read IFC data, automatically generating computational meshes aligned with real architectural layouts.
- **Zone data** (compartments, openings, ventilation nodes) can be read directly, ensuring spatial fidelity between the design model and the CFD domain.

This automation ensures that the CFD model becomes a **true digital twin of the design geometry**, maintaining consistency and traceability.

6.3.3 Cloud-Based CFD Execution

Modern CFD solvers are computationally demanding. Integration with cloud computing platforms — such as Autodesk Construction Cloud or university-based compute clusters — now allows simulations to run remotely, linked directly to the BIM project.

A fire engineer can:

1. Trigger a CFD run directly from the BIM interface.
2. Receive notification when results are available.
3. View output data (temperature, velocity, visibility, CO concentration) as interactive 3D overlays within the BIM viewer.

This workflow transforms CFD analysis from an isolated task into an **embedded design function**.

6.4 Evacuation Modelling Integration

Evacuation simulators — including **Pathfinder**, **BuildingEXODUS**, and **Simulex** — model occupant behaviour, route choice, and travel times to evaluate evacuation performance under fire scenarios.

6.4.1 Data Interoperability

- Building geometry and occupant parameters (stairs, exits, occupancy density) are exported from BIM models using IFC or DXF files.
- Results — egress times, flow rates, congestion points — can be reimported into the BIM model as colour maps or animated sequences.
- Integration with **Smartfire** or **FDS** allows coupling of smoke visibility and temperature data to occupant tenability analysis.

This linkage creates a **combined smoke–evacuation model**, forming a central analytical layer of the Digital Fire Strategy.

6.4.2 Scenario Management

Each evacuation scenario can be stored as a discrete data object within the CDE:

- Scenario ID (linked to Gateway reference)
- Input parameters (occupant load, mobility profile, exit capacity)
- Performance criteria (maximum RSET, target safety margin)
- Results dataset (CSV or JSON files)
- Visual outputs (animations, tenability charts)

These scenario records can be cross-referenced to the main fire strategy dataset, ensuring full auditability.

6.5 Integration Workflow

The workflow for integrating CFD and evacuation modelling within a Digital Fire Strategy typically follows these steps:

Stage	Action	Output / Integration
1	Export building geometry from BIM (IFC/gbXML)	Input file for CFD and evacuation model
2	Define fire scenarios (locations, HRR, ventilation)	Simulation cases logged with unique IDs
3	Run CFD simulation (Smartfire/FDS)	Temperature, smoke, and visibility datasets
4	Import results into BIM model (visual overlays)	Smoke layer maps, temperature contours
5	Run evacuation simulation (Pathfinder/EXODUS)	RSET, flow rate, and congestion data
6	Integrate CFD and evacuation outputs	ASET–RSET comparison, performance report
7	Upload datasets to CDE and link to Digital Fire Strategy container	Verified analytical evidence accessible via Golden Thread

This integrated workflow ensures analytical transparency and allows reviewers to interrogate both the model geometry and the simulation outcomes within the same digital ecosystem.

6.6 Structuring Modelling Data

To make simulation results usable within a Digital Fire Strategy, they must be **structured** and **machine-readable**. The following schema is commonly adopted:

Data Category	Format	Description
Input Parameters	JSON / XML	Fire load, HRR curve, boundary conditions
Geometry	IFC / gbXML	Spatial representation of compartments
Results	CSV / HDF5	Time-series of temperature, visibility, CO concentration
Visualisation	PNG / MP4 / WebGL	Animations and contour plots
Validation Data	PDF / CSV	Comparison against test or benchmark results

Each dataset is linked to a **unique identifier** corresponding to the relevant fire strategy section or system element. This structure allows simulation data to be archived, retrieved, and verified as part of the Golden Thread record.

6.7 Validation and Regulatory Acceptance

Regulators increasingly recognise the value of digital modelling but expect clear documentation of methodology and assumptions.

For model-based evidence to be accepted:

- The **software version**, **solver settings**, and **input parameters** must be fully documented.
- The engineer must demonstrate **competence** in using the tools, referencing BS 7974 and CIBSE Guide E.
- Simulations should be **calibrated or benchmarked** against experimental or literature data.
- Results should be presented in both **visual** and **quantitative** form, with clear reference to performance criteria (ASET/RSET, temperature thresholds, toxicity limits).

By embedding this documentation digitally within the CDE, the regulatory review process becomes faster and more transparent.

6.8 Live Data Feedback and the Digital Twin

The next evolution of CFD and evacuation modelling is their integration within **Digital Twin** platforms. Here, models are not static but continuously updated using live sensor data — such as temperature, smoke concentration, or door status.

- CFD models can adjust boundary conditions dynamically to simulate real-time fire development.
- Evacuation models can use live occupancy data (from sensors or access control systems) to predict evacuation times under current conditions.
- These models can inform automated alerts, system optimisation, and emergency response planning.

In this context, the Digital Fire Strategy becomes a **predictive tool**, capable of anticipating risks and guiding operational decisions rather than merely documenting compliance.

6.9 Cloud Collaboration and Version Control

Simulation models and results are stored and versioned within the **Common Data Environment**:

- Each simulation has a **version tag** (e.g., CFD_V03_2025-09-15).
- Changes in geometry, fire load, or system design automatically trigger an update.
- Reviewers can access previous runs for comparison and validation.

This provides a **continuous analytical audit trail** — a record of how the fire safety design evolved and how performance was demonstrated at each stage.

6.10 Case Example: High-Rise Residential Development

In a 30-storey residential tower, the Digital Fire Strategy was linked to the project's Revit model via BIM 360. Key steps included:

- **Smartfire** imported the full building geometry through IFC.
- Fire scenarios were defined for four representative floors.
- CFD results (smoke movement, temperature, visibility) were exported to Revit as volumetric overlays.
- **Pathfinder** evacuation simulations were run using the same geometry, with results showing total evacuation within RSET < ASET.
- The combined results were uploaded to the CDE and linked to the fire strategy's Section 5.3 "Means of Escape – Performance Validation."

The BSR reviewer was able to open the BIM model and view the simulations directly — eliminating the need for separate PDF reports.

This case demonstrates the **practical realisation of digital performance-based fire design**.

6.11 Future Developments

The integration of performance modelling into Digital Fire Strategies will continue to advance, with several trends emerging:

- **AI-Assisted Modelling:** Automated scenario generation and convergence testing using machine learning.
- **Parametric Design Integration:** Linking fire performance metrics (e.g., ASET/RSET) to design variables for optimisation.
- **Unified Data Standards:** Expansion of IFC property sets to include simulation metadata and performance outcomes.
- **Regulatory Portals:** Direct digital submission of modelling data for automated review and archiving within the Golden Thread.

These innovations will make simulation not a specialist task, but a **standard component of digital design validation**.

6.12 Summary

The integration of **CFD, evacuation, and performance modelling** into BIM and CDE systems marks a fundamental advancement in fire safety engineering.

Where once simulations were isolated technical exercises, they are now part of a **continuous digital feedback loop** — linking analysis, design, and operational safety in one ecosystem.

Key outcomes include:

- Automated geometry exchange between BIM and simulation tools.
- Structured data storage for validation and reuse.
- Transparent, auditable evidence for regulators and clients.
- The emergence of predictive modelling through digital twins.

As digital fire strategies mature, computational modelling will become both a **design verification method** and a **live safety management function** — a bridge between engineered prediction and operational performance.

Chapter 7
Digital Validation and Automation

7.1 Introduction

The traditional process of verifying a fire strategy has long relied on manual checking — the fire engineer reviews drawings, compares design elements to code requirements, and documents compliance through narrative justification. While effective, this method is time-consuming, susceptible to oversight, and poorly suited to projects with thousands of fire-critical components.

In the digital era, **validation and automation** are redefining this process. By embedding rules, logic, and artificial intelligence into digital models, it becomes possible to **automatically test** whether fire safety provisions meet prescriptive or performance-based criteria.

The **Digital Fire Strategy**, when structured correctly, acts as the information backbone of this process. It enables rule-based checking, automated notifications of non-compliance, and continuous verification throughout the design, construction, and operational phases.

7.2 From Manual Checking to Digital Verification

Manual checking typically involves multiple documents — fire strategy reports, drawings, specifications, and calculations. As projects grow in scale and complexity, manual cross-referencing becomes both inefficient and prone to inconsistency.

Digital verification replaces this with **automated consistency checks**, where the model itself becomes the test environment.

In this process:

1. The fire safety data is structured within the BIM model (see Chapter 4).
2. Software interprets design rules (e.g., escape distance limits, fire rating requirements).
3. The system automatically identifies elements that do not conform.

This marks a shift from **reactive review** to **proactive assurance**, enabling fire safety issues to be resolved early and consistently.

7.3 Foundations of Automated Rule Checking

Automated rule checking is built upon three essential components:

1. **Structured Data** – Each element must contain standardised parameters such as fire rating, door width, or distance to nearest exit.
2. **Digital Rulesets** – Codified versions of design standards (e.g., Approved Document B, BS 9999, BS 7974).
3. **Validation Engine** – Software that evaluates the data against the ruleset and generates compliance results.

These components work together within BIM or CDE environments to perform **machine-readable validation** of the fire safety design.

7.4 Digital Rulesets and Standards Encoding

Fire safety standards can be encoded into machine-readable form, enabling automated interpretation. For example:

- **Approved Document B (ADB)** – defines limits for compartment size, travel distances, and escape route widths.
- **BS 9999** – provides performance-based risk profiles and design flexibility parameters.
- **BS 7974** – sets out the framework for engineering-based analysis and assessment.

Digital rulesets translate these requirements into logical statements such as:

- *"If occupancy type = Residential and height > 11 m, then provide sprinklers in accordance with BS 9251."*
- *"If door serves protected route, then FireRating ≥ 30 min and SmokeControl = True."*
- *"If travelDistance > 18 m (unsprinklered), flag as non-compliant."*

By embedding these conditions in software, compliance can be tested automatically and consistently across the entire model.

7.5 Rule-Based Validation Tools

Several software tools already support automated rule checking and model validation:

Tool / Platform	Primary Function	Fire Strategy Application
Solibri Model Checker	Rule-based model validation using predefined or custom rule sets	Verification of fire ratings, travel distances, protected shafts
Navisworks Manage	Clash detection and visual coordination	Identifies breaches of compartmentation and system conflicts
dRofus	Data management and validation tool integrated with Revit	Tracks fire safety attributes and performs parameter consistency checks
Autodesk ACC / Forge API	Cloud-based rule automation	Custom scripts for real-time model compliance monitoring
BIMCollab / BIMTrack	Issue tracking linked to model validation	Logs non-compliance items directly to CDE for audit trail

These platforms allow fire engineers to set up automated workflows where validation occurs continuously, not just at review milestones.

7.6 Example: Automated Compliance Workflow

An automated digital validation process for a fire strategy could follow this workflow:

1. **Model Preparation**
   a. All fire-related elements (walls, doors, compartments) tagged with parameters (fire rating, smoke control, etc.).
2. **Ruleset Selection**
   a. Select or configure digital ruleset (e.g., "Residential Building – ADB Vol. 1").
3. **Model Analysis**
   a. Software checks each element against criteria: travel distance, fire door ratings, stair widths.
4. **Results Generation**
   a. Outputs a report listing compliant, non-compliant, and unverified elements.
   b. Visual markers applied in the 3D model (e.g., red highlight for non-compliance).
5. **Issue Resolution**
   a. Engineer or designer corrects model; issue status automatically updates.
6. **Regulatory Audit Trail**
   a. Results stored in the CDE, version-controlled, and linked to Gateway submission records.

This creates a **closed-loop validation cycle** ensuring continuous compliance and full traceability.

7.7 AI and Machine Learning Applications

Artificial Intelligence (AI) and Machine Learning (ML) are now being applied to enhance digital validation. Their primary advantages include **pattern recognition**, **prediction**, and **adaptive learning** — allowing systems to identify trends and refine checks based on previous project data.

Applications include:

- **Design Pattern Recognition:** AI analyses historical compliant designs to predict likely compliance issues in new models.
- **Automated Code Updates:** AI updates rule sets automatically when standards change.
- **Predictive Risk Scoring:** Algorithms estimate the probability of non-compliance based on project type, complexity, and design parameters.
- **Natural Language Processing (NLP):** AI reads narrative strategy text and cross-references it with model data to ensure consistency.

Over time, AI will enable fire strategy validation systems to move beyond binary compliance and toward **probabilistic assurance**, where design confidence is quantified and continuously refined.

7.8 Linking Validation to the Golden Thread

Digital validation is inseparable from the **Golden Thread** principle of information integrity. Every validation report, rule check, and issue resolution must be **recorded and version-controlled** within the CDE, ensuring:

- Full transparency of design decisions.
- Clear accountability for who resolved each issue.
- A historical log of how the fire strategy evolved from concept to completion.

When linked to Gateway submissions, this provides regulators with a **complete audit trail** — evidence that every element of fire safety was validated, tracked, and approved through controlled digital workflows.

7.9 Continuous Compliance During Operation

Once a building is occupied, validation does not end. The Digital Fire Strategy can continue to operate as a **compliance monitoring system**, integrating with CAFM and Digital Twin platforms.

For example:

- A fire door's maintenance record is uploaded to the CAFM system.
- The CDE automatically validates whether the door still meets its rated performance (e.g., certificate in date, condition acceptable).
- If non-compliance is detected, the system alerts the Building Safety Manager.

This enables **continuous assurance** — ensuring that the operational fire strategy remains accurate, auditable, and live throughout the building lifecycle.

7.10 Case Example: Automated Checking in a Healthcare Facility

A hospital redevelopment project implemented automated validation for fire safety within Autodesk BIM 360. Fire doors, dampers, and walls were tagged with attributes consistent with BS 8644-1. Solibri was used to apply rule checks such as:

- Fire doors in escape routes must have ≥30-minute rating.
- Duct penetrations through compartment walls must include fire dampers.
- Protected lobbies must maintain minimum width and direct escape path.

The system identified 147 issues early in design, of which 96% were resolved before Gateway 2 submission. The regulator approved the digital submission within three weeks — half the usual review time — demonstrating the **efficiency and transparency** of automated compliance checking.

7.11 Benefits and Limitations

Benefits

- **Speed:** Instant compliance checks across thousands of elements.
- **Consistency:** Eliminates subjective interpretation of repetitive rules.
- **Transparency:** Provides a verifiable audit trail of all decisions.
- **Early Detection:** Identifies issues before construction begins.
- **Regulatory Confidence:** Facilitates faster, evidence-based approvals.

Limitations

- **Rule Complexity:** Not all standards are easily codified; performance-based assessments still require engineering judgement.
- **Data Quality Dependence:** Validation is only as accurate as the data within the model.
- **AI Transparency:** Machine learning models must remain explainable to ensure regulatory trust.
- **Interoperability:** Varying BIM formats and software versions can affect rule application.

The key is to view automation as **an augmentation of professional competence**, not a replacement for engineering judgement.

7.12 Future Directions

The next decade will likely see:

- **Unified Fire Safety Rules Libraries** maintained by standards bodies (e.g., BSI or CIBSE).
- **Automated Gateway Submission Systems** linked directly to BSR review portals.
- **Real-Time Validation Dashboards** within BIM viewers, showing live compliance status.

- **Integration with Digital Twins** for continuous feedback on in-use performance.

Ultimately, automated validation will underpin a **data-driven regulatory ecosystem**, where fire safety assurance is demonstrable, dynamic, and fully integrated into the design and operation of buildings.

7.13 Summary

Digital validation and automation mark a pivotal step in the evolution of fire strategy practice.

By embedding logic and intelligence within digital models, compliance can be tested continuously, accurately, and transparently.

Key outcomes include:

- Consistent, rule-based verification against national standards.
- Reduced design risk and faster regulatory approval.
- Enhanced collaboration and traceability through the Golden Thread.
- A pathway toward intelligent, self-verifying fire safety systems.

The **Digital Fire Strategy** thus becomes not only a record of design intent but a *living system of assurance* — capable of self-checking, adapting to change, and maintaining integrity throughout the building lifecycle.

Chapter 8
Digital Handover and the Golden Thread

8.1 Introduction

The culmination of every fire safety design process is the **handover of information**. In traditional projects, this occurred as a collection of static documents — drawings, certificates, and reports — handed to the client at practical completion. Yet these documents often became outdated almost immediately, with their updates lost between design teams, contractors, and facilities managers.

The digital era replaces this fragmented approach with a **continuous data flow**, ensuring that the fire strategy and all related safety information remain live, traceable, and accessible throughout the building's lifecycle. This concept — known as the **Golden Thread of Information** — lies at the heart of the **Building Safety Act 2022**.

The **Digital Fire Strategy** is the central component of that Golden Thread. It is the structured, validated, and version-controlled dataset that connects the design intent to the operational reality of the building.

8.2 Understanding the Golden Thread

The Golden Thread represents a **single, enduring source of truth** for building safety information. It ensures that anyone responsible for safety — from designers to building operators — can access accurate and current data at any time.

The Golden Thread must be:

- **Accurate:** Reflecting the latest verified design and installation information.

- **Accessible:** Readily available to those with authorised access, including regulators.
- **Accountable:** Showing clear ownership and responsibility for each dataset.
- **Auditable:** Maintaining version history and change logs.
- **Secure:** Protected from unauthorised modification or misuse.

For the fire engineer, the Golden Thread transforms the fire strategy from a design deliverable into a **long-term safety record**, integral to building management and compliance.

8.3 Legal Foundations

Under the **Building Safety Act 2022**, dutyholders — including the **Principal Designer**, **Principal Contractor**, and **Accountable Person** — are legally responsible for maintaining the integrity of the Golden Thread.

Key Provisions:

- The Golden Thread must contain all information needed to **understand and manage building safety risks**.
- It must be **stored digitally** in a format that allows structured retrieval and updating.
- It must be **maintained through the lifecycle** of the building — from design to occupation and refurbishment.
- The **Accountable Person** must ensure its accuracy and accessibility to the **Building Safety Regulator (BSR)** upon request.

Failure to maintain accurate digital safety records can lead to regulatory enforcement, including stop notices and penalties.

8.4 The Digital Fire Strategy as the Core of the Golden Thread

The **Digital Fire Strategy (DFS)** forms the backbone of the Golden Thread, integrating data from all safety-critical systems.

It includes:

- Fire safety design intent and performance objectives.
- Validated data from BIM, CFD, and evacuation models.
- Product and material certifications.
- Inspection, testing, and maintenance records.
- Version-controlled updates reflecting as-built and as-managed conditions.

By linking these datasets through unique identifiers (as defined in **BS 8644-1**), the DFS becomes a **living reference document** — constantly aligned with the real building and its operational systems.

8.5 Information Containers at Handover

At project completion, information must be organised into structured **digital containers**, consistent with BS 8644-1 and ISO 19650 principles.

Typical fire safety information containers include:

Container Name	Contents	Purpose
Fire Strategy Model (FSM)	As-built BIM model with verified fire safety attributes	Defines final fire strategy geometry and systems
Fire Safety Record (FSR)	Installation data, commissioning certificates, and test results	Confirms performance of installed systems
Digital Fire Safety File (DFSF)	Consolidated record of all validated fire safety data	Provides the operational baseline for the Accountable Person
Change Log Register (CLR)	Documented record of revisions and authorisations	Maintains traceability and version control

These containers are stored within the **Common Data Environment (CDE)** and linked to other safety information — structural design, M&E systems, and maintenance schedules — to form a complete digital safety record.

8.6 The Handover Process

Step 1: Verification

Before handover, the fire engineer verifies that:

- All fire strategy data is up to date.
- Fire-resisting elements and systems match the as-built model.
- Validation and testing results are approved.
- Documentation (certificates, inspection records) is digitally linked.

Step 2: Approval and Submission

The verified datasets are submitted to the **Principal Designer** and **Principal Contractor** for joint approval.

Gateway 3 submission to the BSR requires confirmation that the fire strategy and safety information meet the requirements of Regulation 38 and the Building Safety Act.

Step 3: Digital Transfer

Upon approval, the DFS and associated containers are transferred to the **Accountable Person** through the CDE.

Each dataset is version-locked, ensuring the handover represents a frozen, verified baseline.

Step 4: Post-Handover Management

The Accountable Person maintains the DFS, updating it to reflect:

- Maintenance and inspection outcomes.
- Alterations or refurbishments.
- System upgrades or decommissioning.

This ensures **continuity of accuracy** from completion into operation.

8.7 Integration with CAFM and Asset Management Systems

Following handover, the Digital Fire Strategy connects directly to **Computer-Aided Facilities Management (CAFM)** platforms and **asset registers**.

For example:

- Each fire door in the BIM model corresponds to a record in the CAFM system.
- Inspection outcomes update the same object's attributes in the DFS.
- Maintenance schedules and fault logs are synchronised automatically.

This integration maintains a **real-time link** between operational management and design intent, closing the loop between **performance verification** and **ongoing compliance**.

8.8 Change Management and Version Control

The Golden Thread requires that all modifications are traceable.

Any change to the building — whether a door replacement, system upgrade, or layout alteration — must be:

- **Proposed** with justification.
- **Approved** by a competent person (e.g., Fire Engineer).
- **Implemented** and recorded in the DFS.
- **Versioned** with timestamp, author, and reason for change.

The CDE automatically manages this process, maintaining an immutable log that demonstrates compliance history — critical evidence during regulatory audits or post-incident investigations.

8.9 The Role of the Accountable Person

The **Accountable Person (AP)** is legally responsible for ensuring that the fire safety information remains accurate and complete.

Their key duties include:

- Maintaining the Golden Thread and updating it after any modification.
- Ensuring that safety information is accessible to residents and regulators.
- Engaging competent professionals (e.g., Fire Engineers) for updates or reviews.
- Coordinating with the **Building Safety Manager** to act on inspection findings.

In essence, the AP becomes the **long-term custodian** of the Digital Fire Strategy, ensuring that design intent, operational performance, and maintenance records never diverge.

8.10 Regulatory Gateway 3 and Digital Submission

Gateway 3, the final regulatory checkpoint, confirms that:

- The constructed building matches the approved design.
- Safety-critical systems are installed, tested, and functioning.
- The Golden Thread is complete and accessible.

The Digital Fire Strategy forms the core evidence submission for Gateway 3. Typically, the digital submission includes:

- Verified BIM model (IFC).
- Digital Fire Safety File (structured XML / PDF).
- Validation and commissioning records.
- Statement of compliance by the fire engineer.
- Gateway 3 declaration from the Principal Designer and Contractor.

This submission ensures that the regulator can review and approve fire safety performance directly from the digital record.

8.11 Operational Continuity and Digital Twins

After occupation, the validated Digital Fire Strategy integrates into the **building's digital twin**, allowing continuous monitoring of safety systems through live data feeds.

Examples:

- **Fire alarm systems** report testing results automatically into the DFS.
- **Smoke control systems** update operational status via IoT sensors.
- **Fire doors** equipped with RFID or QR tags link inspection data to model elements.

These live updates allow the Accountable Person to confirm compliance dynamically, replacing static fire safety files with an **active, self-updating system**.

8.12 Access and Transparency

The Golden Thread promotes transparency between building owners, residents, and regulators.

Digital platforms can provide role-based access to safety information:

- **Residents:** Summary dashboards showing key fire safety measures.
- **Regulators:** Full technical datasets and audit trails.
- **Maintenance Teams:** Interactive maps showing system locations and maintenance intervals.

This openness fosters **trust and accountability**, aligning with the post-Grenfell principles of safety culture reform.

8.13 Long-Term Data Retention

To maintain the Golden Thread over the life of the building, data must be:

- **Retained securely** for the building's full operational period.
- **Migrated** when software or platforms evolve, avoiding obsolescence.

- **Backed up and archived** in compliance with organisational policies.

Periodic data audits should confirm:

- Data integrity and accessibility.
- Compliance with privacy and cybersecurity standards.
- Alignment between design intent and current building configuration.

These measures ensure that the Digital Fire Strategy remains a **trusted, enduring record** — not a transient digital artefact.

8.14 Summary

Digital handover represents the pivotal link between design, construction, and operation.

The **Digital Fire Strategy** is the central mechanism by which the Golden Thread is created, maintained, and applied — ensuring that fire safety intent remains transparent and verifiable across the entire building lifecycle.

Through structured data, controlled workflows, and secure management systems, the Golden Thread delivers:

- A permanent digital record of fire safety design and verification.
- Clear accountability across dutyholders.
- Real-time integration with operational management systems.
- Ongoing confidence for regulators, occupants, and owners alike.

The result is a shift from project completion as an endpoint to **project continuity** — where safety, information, and accountability remain connected for the life of the building.

Chapter 9
CAFM Integration and Maintenance Records

9.1 Introduction

Once a building is occupied, the responsibility for fire safety moves from the design and construction team to the **Accountable Person** and their operational management system.

At this stage, the fire strategy ceases to be a static report and becomes a **living operational tool** — informing inspection regimes, guiding maintenance activities, and supporting regulatory compliance.

Integration between the **Digital Fire Strategy (DFS)** and **Computer-Aided Facilities Management (CAFM)** systems ensures that safety-critical data is not only preserved but also **actively used**. This chapter explores how that integration works in practice, the types of data involved, and the processes that maintain the **Golden Thread** in operational environments.

9.2 The Role of CAFM in Fire Safety Management

A **CAFM system** is a digital platform used by facilities managers to plan, track, and record building maintenance and operations.

When linked with the Digital Fire Strategy, CAFM systems become the operational interface of the Golden Thread — ensuring that every asset linked to fire safety is monitored, maintained, and recorded.

Typical CAFM platforms include:

- **Planon**
- **Concept Evolution**
- **FSI Concept / MRI Evolution**
- **Service Works Global (SWG)**

- **Archibus**
- **Maximo**

These systems store maintenance schedules, inspection logs, and compliance records for thousands of assets — from fire doors and dampers to detectors, sprinklers, and extinguishers. Integration with the DFS ensures that these assets are not managed in isolation but remain directly linked to the **fire safety design intent** defined in the original model.

9.3 Linking the Digital Fire Strategy to CAFM

Integration between the DFS and CAFM platform is typically achieved through **data exchange protocols** such as:

- **IFC (Industry Foundation Classes)** – for model-based asset transfer.
- **COBie (Construction Operations Building information exchange)** – for structured asset and maintenance data.
- **API (Application Programming Interface)** – for live data synchronisation between systems.

During handover (see Chapter 8), the fire strategy model is exported into a format compatible with the CAFM system. Each fire-safety-critical component — a door, damper, alarm, or riser — becomes a **linked asset record** within the CAFM environment, carrying attributes such as:

Attribute	Example Value
Asset ID	FD-L3-072
System	Fire Door – Lobby Level 3
Fire Resistance	E30 / Sa
Certification Reference	IFCC 2478
Maintenance Interval	6 Months
Responsible Person	Building Safety Manager
Last Inspection	2025-08-15
Condition	Serviceable
Linked Model GUID	4D3F-8B21-CC97-0021

These attributes ensure that the digital representation and the physical asset remain **synchronised** throughout the building's life.

9.4 Maintenance Record Management

Maintenance records form the operational backbone of the Golden Thread. Every inspection, repair, or replacement event must be **digitally recorded**, forming a traceable history for each asset.

Typical records include:

- **Scheduled inspections:** Fire doors, dampers, alarms, and extinguishers.
- **Reactive maintenance:** Fault reports and emergency repairs.
- **Certification and testing:** Annual service certificates, test results, and installer sign-offs.
- **Component replacement:** Updated product data and installation evidence.

Each record is **time-stamped**, linked to the responsible engineer, and automatically cross-referenced to the relevant fire strategy section (e.g., *Fire Protection Installations*).

This creates a **continuous feedback loop** between operational maintenance and design intent — ensuring that any deviation or deficiency can be immediately identified and corrected.

9.5 Example: Fire Door Asset Lifecycle

The lifecycle of a single fire door illustrates the integration process clearly:

1. **Design Stage** – Fire door specified in the BIM model with 30-minute rating and smoke seal attributes.
2. **Construction Stage** – Door installed, tagged with unique QR or RFID code, and verified in as-built model.
3. **Handover** – Door data exported to CAFM system, forming part of the operational asset register.

4. **Operation** – CAFM system schedules biannual inspection. Maintenance contractor scans QR tag and records condition.
5. **Maintenance Update** – If defect found, replacement logged; CAFM updates door record and pushes update to DFS.
6. **Regulatory Audit** – Regulator views record in CDE, confirming maintenance and compliance history.

This process ensures that every asset remains traceable from design through operation — a **single line of accountability** enabled by digital integration.

9.6 Digital Inspection and Mobile Data Capture

Modern CAFM systems support **mobile inspection applications**, allowing maintenance teams to conduct inspections using tablets or smartphones. Technologies such as **QR codes**, **RFID tags**, and **Bluetooth sensors** enable field engineers to access and update fire safety data instantly.

Typical process:

1. Engineer scans asset tag.
2. Mobile app retrieves latest DFS data from cloud database.
3. Inspection checklist completed and uploaded with photos.
4. Non-compliance items automatically generate tasks or alerts.
5. Updated data synchronises to the DFS and CDE.

This approach ensures immediate update of safety records, eliminating delays associated with manual reporting or spreadsheet uploads.

9.7 Data Flow and Synchronisation

A robust data flow between the DFS, CAFM, and CDE ensures that information remains consistent across systems.

Example Data Flow:

1. **BIM Model → CAFM:** Export of verified as-built asset data.

2. **CAFM → CDE:** Ongoing upload of maintenance and inspection records.
3. **CDE → DFS:** Periodic synchronisation and versioning of updated fire safety data.
4. **DFS → Regulatory Portal:** Read-only access for auditors or the BSR.

APIs or middleware (such as COBie translators or Autodesk Tandem connectors) can automate this exchange, maintaining **data fidelity** between design and operation.

9.8 The Role of the Building Safety Manager

The **Building Safety Manager (BSM)** plays a key role in maintaining the operational integrity of the DFS and CAFM link.

Their duties include:

- Scheduling and overseeing inspections.
- Ensuring contractors record accurate data.
- Monitoring system alerts and fault reports.
- Liaising with the fire engineer for technical interpretation.
- Preparing annual compliance statements for the Accountable Person.

The BSM acts as the **operational custodian** of the Golden Thread, ensuring that fire safety systems perform as designed and that data remains current and reliable.

9.9 Predictive Maintenance and Analytics

The integration of CAFM with IoT sensors and analytics platforms is enabling a move toward **predictive maintenance** — identifying potential failures before they occur.

Examples:

- **Door sensors** detect excessive opening frequency, indicating wear on closers or seals.

- **Sprinkler flow meters** detect partial blockages through abnormal readings.
- **Smoke control fans** monitored for performance drift or vibration anomalies.

AI algorithms analyse these data streams to flag early warnings. Such predictive insights can automatically trigger CAFM work orders and update risk assessments within the DFS, creating a truly **proactive fire safety management system**.

9.10 Integration with Digital Twins

The **Digital Twin**, discussed in Chapter 10, represents the next stage of this evolution.

When the DFS and CAFM data feed directly into a live digital model:

- System status (alarms, detectors, dampers) is visualised in real time.
- Maintenance status and performance trends appear as dynamic dashboards.
- Occupant and sensor data feed back into ongoing risk assessments.

This integration provides a **real-time picture** of fire safety performance, turning maintenance data into actionable intelligence.

9.11 Reporting and Regulatory Assurance

Digital CAFM-DFS integration simplifies regulatory reporting and compliance audits. At any time, the Accountable Person or regulator can retrieve:

- Up-to-date maintenance records.
- Evidence of inspection and remedial actions.
- Certificates of compliance linked to specific assets.
- Audit trails showing responsible persons and timestamps.

This removes ambiguity and provides a **single authoritative source** of evidence — aligning with the Golden Thread's principle of transparency and traceability.

9.12 Benefits and Challenges

Benefits	Challenges
Continuous record of fire safety performance	Integration between different software ecosystems
Reduced administrative burden through automation	Data quality control across multiple contractors
Improved response to non-compliance	Cybersecurity and data access management
Predictive maintenance improves reliability	Ongoing training for staff in digital systems
Clear audit trail for regulators	Consistent alignment of DFS and CAFM data structures

Effective governance and data standards mitigate these challenges, ensuring the operational benefits of integration are fully realised.

9.13 Case Example: University Campus Integration

A large UK university integrated its fire strategy dataset with a CAFM system managing over 12,000 assets. Fire alarm zones, extinguishers, and fire doors were digitally tagged and maintained through a single interface.

- Inspection teams used mobile apps to scan and update asset data.
- Automatic alerts were generated when certification lapsed.
- The Digital Fire Strategy model in Autodesk BIM 360 was updated weekly via API synchronisation.

During an external audit, the university demonstrated full traceability from specification to inspection records, meeting all requirements of BS 8644-1 and the Building Safety Act.

The system also identified recurring defects in a specific door type, allowing proactive replacement — a demonstration of **data-driven maintenance** in practice.

9.14 Summary

The integration of the **Digital Fire Strategy** with **CAFM and asset management systems** transforms building safety management from a periodic, manual process into a continuous, data-informed discipline.

Through this integration:

- Every fire-safety asset remains linked to its original design intent.
- Inspection, maintenance, and certification data feed directly into the Golden Thread.
- Compliance can be verified in real time through auditable digital records.
- Predictive analytics enable proactive intervention, reducing risk and cost.

In essence, the operational phase completes the vision of the Digital Fire Strategy — a **living, evolving system** that unites design intelligence, performance verification, and maintenance assurance into one coherent digital framework.

Chapter 10
Digital Twins and Real-Time Fire Safety Management

10.1 Introduction

The digital transformation of fire strategies does not stop at design handover.

The natural evolution of the **Digital Fire Strategy (DFS)** is its integration into a **Digital Twin** — a dynamic, data-driven model that mirrors the physical building in real time.

In this environment, information from sensors, systems, and human interaction continually updates the digital representation of the building.

The result is a **living safety ecosystem**, where fire engineers, building managers, and regulators can monitor performance, predict risks, and respond proactively rather than reactively.

This chapter explores how Digital Twins are developed, how they integrate with fire safety systems, and how they redefine assurance, maintenance, and emergency response in the era of connected buildings.

10.2 Defining the Digital Twin

A **Digital Twin** is a continuously updated digital replica of a physical asset. It combines geometric data (from BIM), operational data (from sensors and IoT devices), and contextual data (from CAFM and DFS sources) to create a unified representation of the building.

In the context of fire safety, a Digital Twin allows:

- Real-time visualisation of fire system status and performance.
- Simulation of potential incident scenarios using live environmental data.

- Integration of inspection, testing, and maintenance information.
- Instant retrieval of the current Digital Fire Strategy.

In effect, the twin becomes the **interface of the Golden Thread**, transforming stored information into actionable intelligence.

10.3 The Evolution from BIM to Digital Twin

The transition from BIM to a Digital Twin follows a clear progression:

Stage	Digital Asset Description	Purpose
BIM Model (Design)	Static 3D model with embedded fire strategy data.	Defines design intent and compliance.
As-Built Model (Handover)	Verified model aligned with constructed reality.	Forms the baseline for operation.
Digital Fire Strategy (Operational)	Structured dataset integrated with CAFM and CDE.	Maintains safety record and maintenance data.
Digital Twin (Live Operation)	Real-time data-driven model reflecting current system state.	Enables monitoring, prediction, and adaptive response.

This final stage closes the loop between design, operation, and feedback — enabling continuous assurance of fire safety performance.

10.4 Data Sources Feeding the Digital Twin

To function effectively, a Digital Twin relies on continuous data inputs from multiple systems, including:

1. **Fire Detection and Alarm Systems** – Real-time alerts, zone activations, and fault logs.
2. **Smoke Control and Ventilation** – Damper status, fan speeds, and airflow readings.

3. **Sprinkler and Suppression Systems** – Valve states, flow pressures, and test cycles.
4. **Environmental Sensors** – Temperature, CO, and particulate monitoring.
5. **Occupancy and Access Systems** – Live headcounts, access control events, and evacuation routes.
6. **Maintenance Data** – Inspection results and outstanding remedial tasks from CAFM systems.
7. **Building Management Systems (BMS)** – Integration of electrical, HVAC, and emergency power systems.

These inputs continuously refresh the Digital Twin, ensuring that its representation of the building's fire safety condition is **current, verifiable, and responsive**.

10.5 System Architecture

The architecture of a Digital Twin for fire safety typically includes the following layers:

Layer	Function
Physical Layer	The real-world building and its fire protection systems.
Sensor & IoT Layer	Devices collecting operational data (detectors, dampers, flow sensors).
Data Integration Layer	APIs, middleware, and message brokers aggregating data streams.
Information Management Layer	CDE or data warehouse storing structured DFS and CAFM data.
Analytics & Modelling Layer	Algorithms analysing data for anomalies, trends, and risk.
User Interface Layer	Dashboards, 3D viewers, and mobile apps for live visualisation.

These layers are interconnected via secure cloud infrastructure, allowing data to flow in real time between systems and users.

10.6 Real-Time Monitoring and Visualisation

Using 3D dashboards built on platforms such as **Autodesk Tandem**, **Bentley iTwin**, or **NavVis IVION**, building operators can visualise the fire safety status of the building in real time.

Key features include:

- **Live equipment maps** highlighting operational status (e.g., green for normal, amber for fault, red for alarm).
- **Interactive overlays** showing smoke control zones, evacuation routes, and compartmentation boundaries.
- **Performance metrics** such as temperature distribution, airflow velocity, or door status updated dynamically.
- **Historical playback** of previous events for forensic analysis.

This visual transparency enhances situational awareness and supports rapid, evidence-based decision-making during both maintenance and emergencies.

10.7 Predictive Analytics and Risk Forecasting

The Digital Twin's analytical engine uses AI and machine learning to identify patterns in sensor data that may indicate emerging risks.

For example:

- A gradual increase in damper actuation times could signal mechanical degradation.
- Repeated nuisance alarms in a specific zone might indicate environmental drift or system interference.
- Trends in occupant movement data could highlight egress blockages or poor route design.

These insights enable **predictive maintenance** and **risk forecasting**, allowing remedial action before safety performance is compromised.

10.8 Incident Response and Simulation

In the event of an incident, the Digital Twin serves as a live command-and-control interface:

- Fire service and building control personnel can view live sensor feeds, smoke progression, and system responses.
- Integration with CFD modules allows **real-time simulation** of smoke spread or temperature profiles based on active conditions.
- Evacuation models (Pathfinder or EXODUS) can be triggered to estimate remaining egress time using current occupancy data.
- Decision-support dashboards display recommended actions, such as system isolation or ventilation reversal.

This real-time situational intelligence significantly enhances emergency response, coordination, and post-incident analysis.

10.9 Data Governance and Cybersecurity

As Digital Twins aggregate vast amounts of sensitive building information, robust data governance and security controls are essential.

Key considerations:

- **Access Control:** Role-based permissions restricting who can view or modify data.
- **Encryption:** Secure transmission and storage of sensor and system data.
- **Audit Logging:** Complete traceability of user actions and data changes.
- **Resilience:** Redundant servers and disaster-recovery strategies to maintain continuity.
- **Compliance:** Alignment with ISO 27001 (Information Security) and GDPR principles.

Cybersecurity forms part of the fire safety ecosystem itself — safeguarding the integrity of the information that protects life.

10.10 Integration with the Golden Thread

The Digital Twin is the operational realisation of the Golden Thread. Every change in the physical environment — from system activation to component replacement — updates the digital record automatically.

Action	Digital Thread Update
Fire alarm triggered	Event logged in CDE with timestamp and zone reference
Damper replaced	New component data uploaded, version incremented
Inspection completed	CAFM record synchronised, DFS attributes refreshed
Fire drill conducted	Evacuation data analysed and stored for audit
System test failed	Alert raised and logged for corrective action

This ensures a **living audit trail** that reflects both design intent and real-world performance — accessible to the Building Safety Regulator at any point.

10.11 AI-Assisted Continuous Assurance

Advanced Digital Twin systems incorporate **AI-assisted assurance engines** capable of continuous compliance checking.

These algorithms evaluate incoming data streams against fire safety thresholds and regulatory standards:

- "Are all critical systems operational?"
- "Do current evacuation routes maintain required travel distances?"
- "Are smoke control fans achieving design flow rates?"
- "Are maintenance intervals within tolerance?"

If deviations occur, the system automatically generates alerts, work orders, or even escalations to responsible persons.

This transforms compliance from a scheduled review activity into an **always-on assurance function**.

10.12 Post-Incident Forensics and Learning

Following a fire event, the Digital Twin provides a complete digital record of what occurred:

- System activations and sensor readings.
- Time-stamped evacuation progress.
- Smoke spread and temperature evolution.
- Maintenance and inspection history.

These datasets allow investigators and engineers to reconstruct events accurately, supporting lessons learned, legal investigations, and system improvements.

This digital transparency represents a significant advancement over traditional post-incident reporting, offering measurable accountability.

10.13 Practical Implementation Example

A commercial office development in Manchester implemented a live Digital Twin linked to its Digital Fire Strategy.

- Sensors monitored temperature, damper status, alarm circuits, and occupancy in real time.
- When a false alarm occurred, the twin displayed the triggered zone within seconds.
- The system automatically compared conditions to CFD benchmarks, confirming no temperature rise.
- The Building Safety Manager received an alert via mobile dashboard, verifying system integrity without physical inspection.

During annual review, the regulator accessed the same dashboard, viewing historical data and test logs through the CDE portal — a demonstration of **continuous, verifiable assurance** in practice.

10.14 Benefits and Challenges

Benefits	Challenges
Real-time visibility of fire safety systems	High initial setup and integration cost
Predictive and proactive risk management	Data interoperability between systems
Rapid incident response and improved safety	Dependence on reliable sensor infrastructure
Continuous compliance monitoring	Cybersecurity management and resilience
Enhanced transparency and accountability	Need for advanced technical competence

While challenges remain, the benefits of real-time visibility and proactive safety far outweigh traditional reactive approaches.

10.15 Summary

The integration of the **Digital Fire Strategy** into a **Digital Twin** represents the pinnacle of digital transformation in fire engineering. It unites design, validation, and operation into a single, intelligent ecosystem — one that continuously monitors, analyses, and optimises building safety.

Key outcomes include:

- **Real-time insight** into system status and performance.
- **Predictive maintenance** through AI-driven analytics.
- **Dynamic response** to incidents and changing risk conditions.
- **Transparent governance** through the living Golden Thread.

In the era of connected buildings, the fire strategy is no longer a static record — it is a **living, learning, and self-assuring system** that underpins resilience, compliance, and occupant safety.

Chapter 11
Future Horizons and Industry Transformation

11.1 Introduction

The transformation of fire safety practice through digitalisation represents one of the most profound changes in the history of the discipline.

Where fire strategies were once static documents, they are now evolving into **living systems of assurance** — interconnected, intelligent, and integrated with every stage of a building's life.

This concluding chapter explores what lies beyond current adoption: the global movement toward data-driven fire safety, the convergence of technologies such as AI and digital twins, and the evolving role of the fire engineer in a world where safety information is dynamic, predictive, and transparent.

11.2 The Global Shift Toward Digital Safety Governance

Across the world, building safety regulation is entering a digital phase. The UK's **Building Safety Act 2022** has set a benchmark for transparency and information management, but similar transformations are emerging internationally:

- **European Union** – The *Digital Product Passport* initiative under the revised Construction Products Regulation (CPR) mandates traceable digital information for all construction materials, including fire-rated components.
- **Singapore** – The *Integrated Digital Delivery (IDD)* framework embeds fire safety models directly within BIM submissions to the Building and Construction Authority (BCA).

- **Australia** – The *Digital Twin Victoria* project integrates fire risk data into state-wide infrastructure models.
- **United States** – The *National BIM Program* and NFPA's *Smart Fire Protection* initiative are exploring digital verification for fire systems and emergency planning.

These developments signal a **global convergence** toward digital safety management — an ecosystem where the **Digital Fire Strategy** is not optional but essential.

11.3 The Next Generation of Standards

The standards landscape will continue to evolve to support digital fire strategies and performance assurance.

Emerging and Forthcoming Standards:

- **BS 8644-2 (in development)** – Expected to extend the BS 8644 framework into digital lifecycle management and operational feedback.
- **CEN/TC 442 BIM Standards** – Expanding IFC property sets to include fire performance, testing data, and maintenance attributes.
- **ISO 23247 Digital Twin Framework** – Providing structure for real-time data synchronisation between physical assets and digital models.
- **CIBSE Digital Engineering Protocols** – Establishing data classification for building services and life-safety systems.
- **OpenBIM Fire Safety Schema (proposed)** – A global collaborative effort to standardise fire safety attributes across platforms.

These frameworks will underpin the international interoperability of digital fire strategies, enabling consistent, machine-readable exchange of fire safety data worldwide.

11.4 Artificial Intelligence and Autonomous Fire Safety Systems

Artificial Intelligence (AI) is set to move from validation and prediction (as explored in Chapter 7 and 10) into **autonomous decision-making** and **adaptive fire safety control**.

Potential Applications:

- **Self-Adaptive Systems:** AI-controlled ventilation and suppression systems that respond dynamically to sensor data, optimising containment and tenability.
- **Predictive Modelling Engines:** Machine learning algorithms continuously retraining on live building data to forecast risk in specific zones or materials.
- **Automated Fire Strategy Generation:** AI-driven design assistants generating initial fire strategy layouts and performance justifications within BIM environments.
- **Regulatory AI Gateways:** Intelligent submission portals automatically reviewing strategies against evolving standards and issuing compliance scores.

While human oversight remains essential, the long-term direction is clear: AI will become a **co-engineer** — continuously analysing, verifying, and optimising fire safety performance in real time.

11.5 Cloud Ecosystems and Interoperability

The future of digital fire strategies lies in **interconnected cloud ecosystems** rather than isolated project environments. These will link:

- **Design platforms (BIM)**
- **Analytical tools (CFD, evacuation, risk assessment)**
- **CDE repositories (BIM 360, Asite, Viewpoint)**
- **Operational systems (CAFM, BMS, Digital Twins)**

Using open standards such as **IFC**, **JSON-API**, and **RESTful architecture**, data will flow seamlessly between systems, eliminating duplication and enabling "one version of the truth." This universal connectivity will allow fire engineers to interact directly with live building data — viewing maintenance results, alarm histories, and inspection records from within the same model used for design.

11.6 Human Factors and Competence Evolution

Technology alone cannot deliver safer buildings — it must be supported by **competence, ethics, and collaboration**.

As digitalisation accelerates, the professional skillset of the fire engineer will expand beyond physical science to include:

Competence Area	Description
Digital Literacy	Understanding data structures, model management, and interoperability.
AI and Analytics	Ability to interpret and audit algorithmic outputs.
Information Governance	Managing digital safety records ethically and securely.
Systems Thinking	Integrating fire engineering principles across disciplines and technologies.
Stakeholder Communication	Translating complex data into actionable insights for clients and regulators.

Institutions such as the **SFPE**, **IFE**, **CIBSE**, and **IFSM** are expected to revise competence frameworks to reflect these new domains, ensuring practitioners can operate confidently within digital ecosystems.

11.7 Education and Professional Development

Universities and training providers will play a pivotal role in preparing future fire engineers for this digital paradigm.

Curricula will likely include:

- BIM-based fire strategy authoring and coordination.

- CFD and evacuation modelling integrated with digital design.
- Data governance and cybersecurity for building safety.
- AI, automation, and ethics in safety-critical systems.

Continuing Professional Development (CPD) will also evolve, offering modular pathways in **Digital Fire Engineering**, **Information Management**, and **Smart Building Technologies**.

11.8 Cultural and Organisational Transformation

The transition to digital fire strategies is not merely technical — it demands a **cultural transformation** across the built environment sector.

Key shifts include:

- From *document ownership* to *data stewardship.*
- From *periodic compliance* to *continuous assurance.*
- From *siloed specialisms* to *interdisciplinary collaboration.*
- From *reactive management* to *predictive performance optimisation.*

These cultural changes redefine success in fire safety — not as "code compliance achieved," but as **risk continuously understood and intelligently managed**.

11.9 The Global Fire Safety Data Network

As data becomes the primary medium of fire safety knowledge, the industry may move toward shared, anonymised **fire safety data networks** — global repositories of validated design, incident, and performance data.

Such networks could:

- Enable benchmarking across building types and regions.
- Support AI models for predictive fire behaviour.
- Inform the refinement of standards and design methods.
- Allow regulators and insurers to base decisions on empirical evidence rather than assumptions.

By contributing to shared datasets, fire engineers will collectively advance the field — turning data into a **global learning resource** for prevention and resilience.

11.10 The Ethical Dimension

Digitalisation brings new ethical challenges:

- Who owns the fire safety data — the engineer, the client, or the building owner?
- How should privacy be protected when occupancy and sensor data are continuously monitored?
- What accountability exists if an AI system influences a safety-critical decision?

Professional ethics must evolve to address these questions, guided by principles of **transparency, accountability, and human oversight**.

Technology may automate compliance, but it must never replace moral responsibility.

11.11 The Role of the Fire Engineer in 2040

By 2040, the fire engineer's role will be unrecognisable from that of the early 21st century.

They will act as:

- **Data custodian** – managing and assuring information integrity across systems.
- **Digital analyst** – interpreting live data and predictive models.
- **Collaborator** – embedded within multidisciplinary design and operational teams.
- **Ethical guardian** – ensuring technology serves life safety and public trust.

In short, the fire engineer of the future will not simply design safe buildings — they will design **safety ecosystems**: intelligent, adaptive, and resilient.

11.12 Conclusion: From Document to Intelligence

The journey from traditional fire strategies to Digital Fire Strategies represents a fundamental paradigm shift — from static documentation to **intelligent, interconnected systems of assurance**.

Through structured data (Chapter 4), BIM integration (Chapter 5), automated validation (Chapter 7), CAFM synchronisation (Chapter 9), and Digital Twin connectivity (Chapter 10), the fire strategy becomes more than a compliance tool.

It becomes the **central nervous system** of building safety — continuously sensing, analysing, and evolving.

The principles of the Golden Thread — accuracy, accessibility, accountability, and assurance — remain timeless.

What changes is how they are achieved: through **data, connectivity, and intelligence** rather than paper, procedure, and periodic review.

As the profession enters this new era, one truth endures:

Fire engineering is not simply about preventing disaster; it is about enabling innovation safely. In a digital world, that safety must itself be digital — traceable, transparent, and alive.

Epilogue

The Living Legacy of Digital Fire Safety

Fire safety has always evolved alongside the buildings it protects. From the hand-drawn plans of the past to today's integrated digital ecosystems, each generation has redefined what it means to design for safety, resilience, and trust.

The emergence of the **Digital Fire Strategy** represents more than technological progress — it is the embodiment of a cultural and professional transformation. It unites data, design, and duty into a single, living framework that ensures safety information remains **accurate, accessible, and accountable** for as long as the building exists.

Where traditional fire strategies documented compliance, digital fire strategies **demonstrate assurance.**

Where once information was handed over, now it is **handed through** — continuously evolving from concept to occupation, sustained by the Golden Thread of information. And where once fire engineering ended at project completion, it now extends into real-time operation, prediction, and learning.

As buildings become smarter, so too must our approach to safety.

The Digital Fire Strategy, underpinned by structured data, automation, and human competence, ensures that safety is not frozen in time but **actively maintained through time** — a constant dialogue between design intent and operational reality.

Key Takeaways

- **Fire strategies are no longer static documents** – they are dynamic, data-driven systems embedded within digital models, continuously validated and auditable.

- **The Golden Thread** establishes an enduring chain of accountability — from design through operation — ensuring safety information remains current, complete, and connected.
- **BIM integration and structured data** create the foundation for interoperability, allowing fire engineers, designers, and facility managers to collaborate within a shared digital ecosystem.
- **Automation and AI** augment professional judgement, accelerating compliance verification and predictive maintenance while maintaining transparency and human oversight.
- **CAFM and Digital Twin integration** extend the fire strategy into real-time management, enabling live monitoring, performance optimisation, and continuous learning.
- **Competence, ethics, and governance** remain essential — digital tools enhance safety only when guided by integrity, accountability, and professional skill.
- **Global harmonisation** of standards and data schemas will enable shared learning, consistent regulation, and the growth of a truly international digital safety culture.

Looking Forward

The next era of fire engineering will be defined not by the reports we write, but by the systems we design to keep them alive.

Digital Fire Strategies mark the beginning of this transformation — an era where fire safety becomes transparent, predictive, and permanently embedded in the built environment's digital fabric.

The profession now stands at a defining moment: to embrace this digital revolution not as an optional upgrade, but as the **new foundation of safety itself**.

Future generations will not ask whether a building complied; they will expect to know **how it continues to comply, perform, and protect** — every day, in real time.

And that is the enduring promise of the Digital Fire Strategy:

A bridge between knowledge and action, design and life, intent and reality — ensuring that the safety we engineer today continues to safeguard tomorrow.

Appendix 1
Glossary of Terms

Term / Acronym	Definition
ADB	*Approved Document B* – UK building regulation guidance for fire safety design.
AI	*Artificial Intelligence* – computer systems capable of performing tasks that normally require human intelligence, such as pattern recognition, reasoning, or prediction.
API	*Application Programming Interface* – a software link that allows data exchange between two applications, such as BIM models and validation tools.
Asset Register	A structured database listing all building components (e.g., fire doors, dampers, alarms) with their locations, specifications, and maintenance histories.
BIM	*Building Information Modelling* – the digital process of creating and managing information across a building's lifecycle.
BMS	*Building Management System* – controls and monitors building services such as HVAC, power, and alarms.
BS 7974	British Standard providing the framework for fire safety engineering design and analysis.
BS 8644-1	Standard defining digital management and exchange of fire safety information across the building lifecycle.
BS 9999	British Standard for fire safety in the design, management, and use of buildings using a risk-based approach.

Term / Acronym	Definition
BSR	*Building Safety Regulator* – UK regulator responsible for enforcing the Building Safety Act 2022.
CAFM	*Computer-Aided Facilities Management* – software used to manage building operations, maintenance, and compliance.
CDE	*Common Data Environment* – a central digital platform for storing, sharing, and managing building information.
CFD	*Computational Fluid Dynamics* – simulation of airflow, smoke, and heat movement during a fire.
COBie	*Construction Operations Building Information Exchange* – a data format for structured asset information handover.
Data Schema	The structured organisation of data fields, relationships, and rules within a digital model.
DFS	*Digital Fire Strategy* – a structured, validated digital record of fire safety design, management, and maintenance data.
Digital Twin	A live digital replica of a physical building, continuously updated via sensors, IoT devices, and operational data.
Gateway 3	Final regulatory stage under the Building Safety Act confirming compliance and readiness for occupation.
Golden Thread	The continuous, accurate, and accessible record of building safety information maintained throughout a building's life.
IFC	*Industry Foundation Classes* – open data model format for exchanging BIM information.
IoT	*Internet of Things* – network of connected sensors and devices transmitting live data from physical assets.

Term / Acronym	Definition
ISO 19650	International standard for managing information over the lifecycle of a built asset using BIM.
Machine Learning	A subset of AI enabling computers to learn patterns from data and improve predictions over time.
Maintenance Record	A documented history of inspections, servicing, and repairs for safety-critical systems.
Model Validation	The process of checking whether data within a BIM model meets specific rules, standards, or performance criteria.
Performance-Based Design	A design approach that uses engineering analysis and quantitative assessment instead of prescriptive code requirements.
PPM	*Planned Preventive Maintenance* – scheduled servicing of assets to prevent failure or degradation.
Regulation 38	UK building regulation requiring transfer of fire safety information to the building owner upon completion.
Rule-Based Checking	Automated model verification using predefined rules derived from standards and codes.
Smartfire / FDS / PyroSim	CFD and fire simulation tools used for modelling smoke movement and fire behaviour.
Validation Engine	Software system that checks BIM data or fire strategy parameters against encoded design rules.
Version Control	The process of tracking changes to data or documents over time to maintain an audit trail.

Appendix 2
Key Standards and Frameworks

Standard / Framework	Scope and Relevance
Building Safety Act 2022	UK legislation mandating the Golden Thread of building safety information and defining dutyholders.
BS 8644-1:2022	Framework for the digital management and exchange of fire safety information across a building's lifecycle.
BS 7974:2019	Fire safety engineering framework for performance-based design.
BS 9999:2017	Code of practice for fire safety in the design, management, and use of buildings.
BS 9251:2021	Fire sprinkler systems for residential and domestic occupancies.
ISO 19650 (Parts 1–3)	International BIM information management standard, defining processes and responsibilities.
NFPA 950 / 951 (US)	Standards for data exchange and IT infrastructure for fire protection and emergency response systems.
CIBSE Digital Engineering Protocols	UK guidance on structuring and managing building services data.
PAS 14191	Information management and security for digital built environment systems.
CEN/TC 442 BIM Standards	European standards committee focusing on digital construction and interoperability.
ISO 23247	Framework for implementing digital twins of physical assets.
Digital Product Passport (EU)	System for traceable digital product data, relevant to fire-rated construction materials.

Appendix 3
Digital Fire Strategy Workflow Example

The Digital Fire Strategy (DFS) operates as a continuous feedback system, linking design, construction, and operation through structured data.

Workflow Summary

1. **Design Stage**
   a. Fire engineer defines performance criteria in BIM model.
   b. All fire safety elements tagged with key parameters (fire rating, compartment ID, protection type).
   c. Initial validation against ADB, BS 9999, and BS 7974 using rule-based checks.
2. **Construction Stage**
   a. Verified components installed; certificates uploaded to the CDE.
   b. Fire strategy model updated to reflect as-built conditions.
   c. Gateway 2 and 3 submissions validated against design data.
3. **Handover Stage**
   a. Digital Fire Strategy exported as structured data (IFC/COBie).
   b. Submitted to Principal Designer, Principal Contractor, and Accountable Person.
   c. Baseline DFS established for ongoing operation.

4. **Operational Stage**
    a. CAFM integration ensures inspection and maintenance records update DFS automatically.
    b. Any changes or replacements trigger version-controlled updates.
    c. Real-time monitoring begins via Digital Twin interface.

Continuous Assurance

a. AI and rule-based systems verify compliance daily.
b. Alerts generated for overdue maintenance or system faults.
c. Regulatory access provided via secure read-only portal.

www.ingramcontent.com/pod-product-compliance
Lightning Source LLC
La Vergne TN
LVHW100331210926
848480LV00003B/1303